DIANA OSBORNE-MCC

Ashland, Washington

May 30, 1999

HORN OF DARKNESS

HORN OF DARKNESS

Rhinos on the Edge

CAROL CUNNINGHAM

JOEL BERGER

New York   Oxford

OXFORD UNIVERSITY PRESS

1997

Oxford University Press

Oxford   New York
Athens   Auckland   Bangkok   Bogotá   Bombay
Buenos Aires   Calcutta   Cape Town   Dar es Salaam
Delhi   Florence   Hong Kong   Istanbul   Karachi
Kuala Lumpur   Madras   Madrid   Melbourne
Mexico City   Nairobi   Paris   Singapore
Taipei   Tokyo   Toronto

and associated companies in
Berlin   Ibadan

Published by Oxford University Press, Inc.
198 Madison Avenue, New York, New York 10016

Oxford is a registered trademark of Oxford University Press

The authors gratefully acknowledge permission to quote Raffi

Library of Congress Cataloging-in-Publication Data
Cunningham, Carol, 1954-
Horn of darkness : rhinos on the edge /
Carol Cunningham and Joel Berger.
p. cm.   Includes bibliographical references and index.
ISBN 0-19-511113-3
1. Black rhinoceros.   2. Wildlife conservation—Africa.
I. Berger, Joel.   II. Title.
QL737.U63C86   1997
599.72'8—dc20   96-22323

9 8 7 6 5 4 3 2 1

Printed in the United States of America
on acid Free Paper

For our parents,
who taught us to value
all living things

Contents

III   YEAR OF THE SCORPION [1993]

IV   YEAR OF THE HUMAN [1994]

A Black Rhino Time Line

1658 Jan van Herwaerden met a "renoster ... die twee hoorens op de neus hadde staen, gelijck de bocken haer hoorens dragen" (rhinoceros with two horns on the nose like antelopes on the head) in southern Africa

1960 60,000 exist throughout Africa

1976 the species is listed as endangered

1983 extinction complete in Sudan, Ethiopia, Somalia

1988 3,500 remain on the continent

1989 Namibia becomes the first country to dehorn rhinos

1991 Zimbabwe and Swaziland dehorn their rhinos

1992 request by southern African countries to legalize the horn trade denied in Kyoto, Japan, during Convention of International Trade in Endangered Species meeting

1993 Zimbabwe's 1991 population estimate of 2,000 revised to 250

1994 Namibia cancels research on dehorning

1996 status in Angola and Mozambique remains unknown

2000 ? ? ?

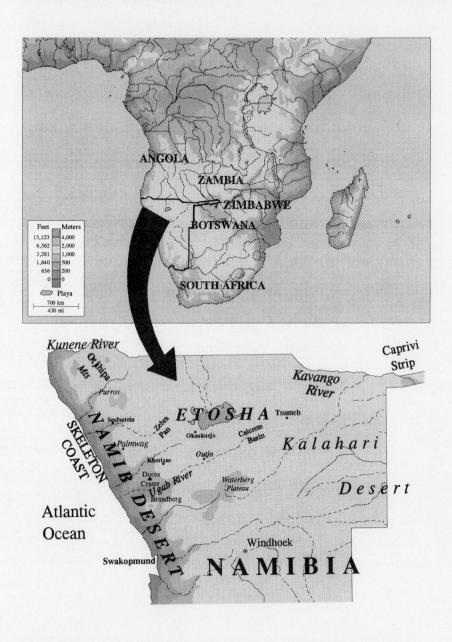

Overview of study areas. *Our* names, Calcrete Basin and Zebra Pan, are fictitious to avoid revealing locations where rhinos still occur and might be poached. Namib Desert names are real, having previously been published by others.

I

Year of the Mopane Fly
[1991]

1

In the Rhino's Path

[Joel]

The black rhino is nature's tank. A charging rhino can gallop at 50 kilometers per hour, chug through dense thornbush, and scatter a herd of elephants. Adult black rhinos have no predators. All animals fear them. Stalking lions will break off a hunt to detour around them.

Black rhinos have been amazingly successful. At the turn of the century, there may have been 100,000 in Africa, scattered from below the Sahara to the Cape. Now, in the entire continent, only one unfenced population of more than 100 animals remains. This remnant roams one of the least hospitable spots on earth—the Namib Desert in northwest Namibia. But even this last refuge is not safe.

Horns are so valuable that people risk their own lives to kill rhinos. Few options remain to stop this deadly harvest. Foot patrols, helicopters, and high-tech solutions have all been tried. All have failed.

A radical strategy to save the rhino has been devised. The logic is straightforward. If a rhino has no horns, the incentive to slaughter it disappears. The solution seems obvious—to cut the rhinos' horns.

* * *

At 10,000 meters above Africa, we looked at each other, two Americans from suburban California, heading to Namibia to track black rhinos. What had we been thinking? How could we put ourselves and our 19-month-old daughter in the path of these animals?

2

Bumbling around in the Bush

[Joel]

The Namib, the oldest desert on earth, is also one of the driest. There, life is a naked quiver tree—the kokerboom—white-barked and thick, clinging to ancient rock; five slender but powerfully built chacma baboons seeking rotten insect larvae; a "white lady" spider hunting scorpions on coastal sands.

Humans and their livestock must also work hard to survive. Goats wander the barren landscape, bulging eyes searching for the last green buds of acacia. In the village of Sesfontein, which sits on a desertified plain, the people are deathly afraid of elephants. When the huge pachyderms visit, they leave the destitute Damara and Herero people with shattered fences, razed crops, and fear. Huts built of saplings, dung, and mud offer no protection. In an area not meant for agriculture, where the annual average rainfall is less than 10 centimeters (4 inches), cherished crops disappearing into mammoth gastric cavities is a tragedy.

One day a man from Sesfontein told us a story. With a tattoo shaped like a dagger gleaming against his dark skin, he described what led to the killing. He blamed despair, alcohol, and a need for money. Once, he had been a proud tracker, a trusted employee of Namibia's elite conservation organization, Save the Rhino Trust. For !Huie, unlike many rural Damaras, there had been options. But these were squandered by the lure of money. Throughout Africa rhinos were being slaughtered. What was different was that !Huie confessed to this killing.

* * *

Windhoek, Namibia – March 18, 1991. It was midnight when we arrived in the silent capital. Hawk moths clung to the warm tarmac. A stag beetle, twice the size of my watch, zoomed past Carol's face, smacking its hardened shell loudly into the floodlight. Our hotel was mercifully quiet, with few signs of such life.

"Goeiemore" (pronounced kwee ya mora), the operator said when we finally answered the phone. We had overslept and now Mrs. Montgomery was waiting in the lobby. Sharon, an ex-journalist working for Save the Rhino Trust, greeted us and took us to her home for a visit. Over thickly brewed Kenyan coffee, we learned about Namibia and its rhinos and politics. The government had just approved a second dehorning operation, one that would begin in two weeks. Security would be tight, but we were welcome to join the operation if we could organize ourselves in time. We decided to try but were confused. Why could a non-government person invite us to the government dehorning?

First on our list were diapers for Sonja. In Windhoek they were called nappies, and, luckily for us, plentiful.

Other problems were not as easily solved. Five of our eight bags, trunks that contained the most critical gear—veterinary rifles, night vision equipment, photographic supplies—hadn't arrived. Neither had containers with radio-electronics and sophisticated rangefinders, or those with our guns. Even my clothes remained somewhere in France. The equipment list that we had supplied to both the U. S. government and the airport authorities was perfectly normal for anyone studying a large, nocturnal mammal. Airport security in Paris thought otherwise. Doesn't the U.S. Central Intelligence Agency work in mysterious ways? Was a diapered baby merely a clever ruse? Why would a family be traveling with sensitive surveillance equipment, radio-electronics, and weapons? So we waited and worried.

Furthermore, the upcoming four-day weekend and the following Easter holidays meant that we had little time to buy a vehicle, obtain temporary residence permits, and meet many of the Namibians with whom we'd work for the next four years. We also needed gear—tools, tents, and maps. A portable refrigerator was vital for storing scientific samples, but finding one and then sorting out how to power it from a car would take days. We hoped Eugene Joubert would help.

It had taken nearly two years to sort out the paperwork and receive official approval for our project. Dr. Joubert had streamlined the mysterious obstacles that had recurrently blocked research clearance. He was the Director of Research for the Ministry of Wildlife, Conservation, and Tourism, and the ex-mayor of Windhoek. He had also studied rhinos in the 1960s—the perfect person to ask for advice.

When we went outside, last night's empty streets were jammed with people. One hundred thousand celebrants crowded the capital for the first anniversary of Namibia's independence from South Africa. Yellow banners were draped across roadways; smiles and laughter filled the streets. Robert Mugabe, the president of Zimbabwe, had arrived for the festivities, and security was tight. Military personnel in camouflage and both uniformed and nonuniformed police lined the boulevards. Guns and armored vehicles were everywhere.

Carol went in search of a phone to call Dr. Joubert, and Sonja and I stood in line at Standard Bank. Four languages, none of them English, rang in our ears; people cut in front of us when we didn't edge forward quickly enough. Soni began screaming and then relieved herself, adding to the already diverse smells. A diaper change was badly needed. *Where was Carol?*

Twenty minutes later she arrived with good news. Dr. Joubert would meet us in an hour at a church less than a kilometer away. We had no idea what he looked like, but he'd be driving a light-colored, Toyota bakkie (small pick-up). *Strange meeting place,* I thought. The banking negotiations dragged on. After 45 minutes, we abruptly arranged to finish our business later. We had to meet Joubert.

We darted through the crowds, a rush of dark faces in darker suits, white legs with black knee socks, and colorful Herero dresses that rustled against the cement sidewalks. Earnest young men with hopeful eyes blocked our path.

"Paper, mister? Paper? English?", they asked, waving newspapers.

Shaking our heads and grunting no, we raced on. Soni bounced like a loose-limbed doll in the baby carrier strapped to my back. Armed soldiers stared as we lurched around people. Then something occurred to us: We couldn't see the church.

Sweat dripped down my sides. We had just come from Nevada's snowy winter; the humidity and heat assaulted us, despite Windhoek's mile-high altitude. My hastily bought clothes abraded my moist skin. We kept moving, slower now, up a sloping, curving street, scanning around and above every wall and gate. At last an ornate, steepled church loomed in front of us. We were four minutes late.

Where was Joubert? What if he'd been early and left? This was certainly the church. "Carol, what exactly did he say?"

"His accent was heavy. I think this is where he said to be." She frowned back at me.

One minute passed, another, then another. We turned toward each vehicle, looking intently, waiting, hoping for a light-colored Toyota. My mind raced. *If we blow this meeting, our first with a high-ranking officer, we'll be in trouble. Big trouble.*

A bakkie turned the corner, a Toyota. We rushed up to it, smiling and gibbering. A bronze-skinned man, middle-aged and perspiring, opened his door.

"Dr. Joubert?" I said in a rush. "Hi. I'm Joel, this is Carol. We're very happy you were able to meet us."

Looking us over, the man quietly apologized and added that he was late for a government meeting. But he'd be glad to help us afterward. Then, like Alice's white rabbit, he disappeared.

We were relieved to have actually met him, but we were also disturbed. He had been polite but not very friendly. And, although he'd said he'd help after his appointment, we didn't know whether to wait, or when his meeting would end. Would he be offended if we left? We felt like idiots. We decided to return to the line at the bank.

Later we called an American couple we knew. As chance would have it, they were having dinner at Joubert's home that night. But then they gave us shocking news: Joubert was offended by our failure to show up at the church. Cunningham and I stared at each other in disbelief. *What?!* What was going on? This was the

Namibian equivalent of the twilight zone. We'd met. We'd talked. He drove a light-colored Toyota bakkie. He told us that he'd help later with our logistics. "Did he ever say his name was Joubert?" we asked each other. As we drove to Joubert's house, we were still confused. Who had we met? It was difficult to stave off depression—banking problems, no luck yet finding a field vehicle, our luggage in Paris, and now a nonmeeting with the ministry's director of research.

Dr. Joubert, or Eugene, as our friends called him, was out golfing when we arrived. A swimming pool shimmered in the late afternoon sun. After we caught up on news, we finally asked our friends what Joubert looked like. "Fiftyish" matched the man we'd met at the church, but the height they mentioned was wrong. Suddenly a man Carol later described as "terribly good-looking" approached. His eyes focused on her. She was trapped. A kiss landed directly on her lips. He turned and spoke to me in perfect, albeit accented, English. We felt mildly hysterical, as Eugene explained why he was late for our meeting at the church. The person we had met was simply an unsuspecting Namibian with the bad luck of having the right type of vehicle. He was just a nice guy who worked for the government and was willing to help. During later supply trips to Windhoek, we'd occasionally spot the "impostor" driving in his light bakkie and yell to each other, "Look, there goes the false Joubert!"

Days passed. Our bags finally came from Paris. Two of the five trunks had been mangled, but the contents were fine. We still needed a vehicle. Range Rovers, Land Rovers, retired United Nations vehicles, dilapidated police vans, jeeps from farms—many old and rusted—burned up our remaining time. We walked the streets, made phone calls, and learned to negotiate Windhoek. But all the vehicles were either too used or cost more than $60,000. Because the dehorning operation would begin in only three days, we had to make a decision. We chose a 1991 Land Rover—the model on display at the dealership.

"The 110 series is very good," assured Luigi Ballotti, the salesman. He seemed to be a direct, honest person. True, he was there to make money, but compared to some of the other salesmen we'd met, he seemed okay.

The cost was far more than we had planned, but we had to have reliability, and something immediately. I was no mechanic, and neither was Carol. We needed a tank, something hefty to carry all our water, petrol, and food for perhaps a month at a time. This particular vehicle had three built-in petrol tanks and a large rack on top. With one cash payment, transportation and sleeping problems were solved. Sonja's bed in the back seat would double for cargo, water jugs, and scientific equipment. I suggested that our new bedroom would be atop the wooden roof rack, reinforced by tubular metal, with lovely black grating. Carol just laughed.

We found a German-made Engel refrigerator. Powered by an extra 12-volt battery that fit snugly under the passenger seat, it would keep film, rhino tissue and fecal samples, milk, and a few fresh groceries cool.

We packed neatly at first: the refrigerator in place, solar panels to one side, then crates and crates of canned food from Woolworth's. We loaded tins of Heel Sampionene (whole mushrooms), Gesnyde Groenboontjies (cut green beans), Kenna kafe (coffee), voorsmelk (fresh milk), irradiated milk for use after the fresh was gone, crackers, and small tins of ham and processed meat mixed with grain. We loaded cases of boxed juices. Large plastic jugs filled with water would sit on the roof rack, along with the king-sized canvas tent. Then came our trunks, sleeping bags, cameras, guns, computer, tarps, and boots. Mosquito nets, a small table, coolers, collapsible chairs, a small upright camping stove, a lantern, towels, detergent, and shade screening, plus another trunk of scientific equipment went in.

It was hopeless. Six 20-liter plastic water containers and six 20-liter metal jerry cans still sat on the ground along with wash basins, clothes, four back packs, a suitcase, two tool boxes, and two first aid kits. Everything had to come. Frustrated, I began shoveling things into the back.

Finally we rolled north. Civilized Windhoek disappeared. We were emancipated from the city's concrete corridors. Carol and I smiled, trying to freeze this moment in our minds. A Jackson Browne cassette played. Our project now had a beginning. Blue skies, jagged mountains, and open country stretched to the horizon. The deep red soils of the western Kalahari caressed the tarmac's edge. Grass, thick stems still green and waist-high, covered the veld. A yellow-billed hornbill flitted across the sky, dipping its wings in wide sweeps. We still needed to master the swaying, boatlike Land Rover and find an air pump, fire extinguisher, and other emergency supplies, but those things didn't matter. We felt good.

Our first destination was Etosha National Park. From there, ministry officials would accompany us to the northern Namib Desert—Damaraland (see glossary), the southern portion of the Kaokoveld, where we'd head for the site of the dehorning operation.

* * *

Black rhinos are one of the most critically endangered species in Africa. The perhaps 100,000 that existed when this century began were scattered from subsaharan deserts to the Cape of Good Hope. They lived on mountains—in thick forest at elevations of up to 3300 meters—and on shadeless gravel plains in Kenya near Lake Turkana. Places that today seem unlikely—Somalia, Ethiopia, the Sudan—harbored black rhinos. White rhinos too were widespread, from the Sudan to South Africa. Like the black rhinos, their numbers have shrunk dramatically. There is only one unfenced population of them left, that in northeastern Zaire's Garamba National Park, and it numbers less than 40. In managed regimes, white rhinos have a more secure future. There are more than 7,000 of them, compared to less than 2,500 blacks. But the world's rhinos—Africa's two species and the three in Asia—are in trouble. With the exception of the white rhino, all are dangerously close to extinction.

Tragically, rhinos are imperilled for one reason: greed. In places like Yemen, China, Korea, Taiwan, and Thailand, their horns are treasured more than gold. In times past, rhino hide was used for sandals, and dung, urine, and blood were used ceremonially, but only the horn is valuable today. Horns are made of a keratinlike substance similar to fingernails. Carved into decorative dagger handles known as jambiyyas, they are worn by men in Yemen, as common as Levis on Americans. And rhino horns had, and continue to have, supposed medicinal qualities in Asia, having been used for at least 4,000 years.

In the past, rhino populations were larger than those today, human populations were smaller, and harvesting by humans then did little to endanger the species. But the situation is now cataclysmically reversed. There are now more than five billion humans and a mere handful of rhinos. In recent years automatic weapons have replaced traps, arrows, and spears, and have given hunters tremendous advantages.

Now facing possible extinction, these hapless creatures are currently the focus of worldwide attention. Millions of dollars have been and are being spent trying to keep them from vanishing altogether. Airplanes, helicopters, and paramilitary operations have been tried. So have monitoring programs, fences and guard dogs, translocations to protected areas, even transoceanic shipments of rhinos to Australia and the United States. Yet populations continue to plummet.

In many ways, rhinos differ dramatically from much of this planet's endangered fauna. They attract disproportionate public support, and not merely from a sympathetic cadre of environmental and conservation groups. The U.S. Congress passed a rhino protection bill in 1994. As distinct members of the mammalian group Perissodactyla, which includes horses, zebras, and tapirs, rhinos have changed little biologically during the past three to four million years. In contrast, most hoofed mammals—the Artiodactyla, which include deer, antelopes, giraffes, and sheep—have been modified while radiating into many of our planet's niches, undoubtedly responding to changing evolutionary pressures. Yet, although rhinos are a flagship species in terms of public recognition, ecologically they are no different from many of the world's large mammals. Whether the greater threat to the species is more of habitat loss because of burgeoning human populations or of poachers killing for lucrative body parts, the result is the same. Populations shrink and are continually fractured into smaller segments, until all the animals are gone.

Impoverishment: It occurs at all levels—biological, economic, and cultural. During the last 100 years, quaggas and bluebuck were driven to extinction by hunting. They were part of a wild Africa that can't be resurrected. Less than 10 years ago, an elephant was killed every eight minutes. Neither animals nor ecosystems operate in a vacuum, not even those protected in national parks. All are subject to market economies, civil strife, and human demography.

The loss of wild rhinos represents more than one type of poverty. Conservationists argue that one sign of a healthy ecosystem is the continuation of ecosystem processes, and there is no doubt that valuable processes are being lost around the globe in an exponentially increasing spiral. But the loss of rhinos may have more immediate and dramatic effects: African communities that could make money from tourism will suffer if these megabeasts die out.

Much of Africa is simply following in the footsteps of the westernized world; as human populations become more urban or expand, wildlife loses. While many of us in the western world struggle to accept the inevitable changes, many Africans are too involved to notice. Staying alive or feeding one's family has, understandably, taken precedence over conservation in places like Somalia, Mozambique, and Angola. Like many people, after having been exposed to such urgent issues, Cunningham and I felt thoroughly stumped trying to work out our individual roles in a world spinning out of control. But we hoped our research would be helpful in the effort to conserve rhinos.

* * *

Getting to Damaraland and the dehorning operation was our immediate concern as we struggled to keep pace with a speeding bakkie. It roared ahead of us, at 110 kilometers per hour, through a monotonous landscape—thick deciduous forests of mopane trees broken only by a few distant hills and kopjes.

At the small community of Khorixas, still 100 kilometers from the Namib's edge, we added 200 liters of petrol to shiny new jerry cans and petrol tanks. A grimy hose, undoubtedly housing oil in a previous life, was used to pour drinking water into six additional containers. Damaras and Hereros gathered to stare at us and our supplies.

Namibian government employees, in fact most Namibians, drive dreadfully fast. We were racing after Dr. Pete Morkel, the veterinarian at Etosha National Park, and Louis Geldenhuys, head of the country's Game Capture Unit. They had led Namibia's first dehorning efforts in 1989 and now planned to remove horns from a different rhino subpopulation. Banking turns too steeply, swaying wildly with each bump, we choked on the enveloping dust. The faster we drove, the more rapidly our supplies were thrust forward into the front seat. Something hard and metallic catapulted from an unknown box and bashed Cunningham in the ear. With each dip in the road our heads flew into the hard ceiling. Soon we learned to thrust a hand up to stave off impact. The Namibian version of off-road cowboys did little to increase our enthusiasm.

"Which way'd they go?"

"Dust, there, turn . . . no, here," we shrieked at each other.

"That's not a road. Go straight. Go straight!"

In thirty minutes the sun would set. We didn't have the faintest idea of where we were. Doros Crater was a tiny blip on an otherwise roadless and undesignated section of the map, somewhere between coastal deserts and inner highlands. Brandberg Mountain, 2,579 meters high, was somewhere to our south, bounded by the foreboding canyons of the Ugab River. The delicate Bushmen paintings and carvings of Twyfelfontein Monument were to the north. We careened somewhere between the landmarks. We wondered about wildlife, not having seen any yet. Ostriches, mountain zebras, even giraffes were supposed to be here. So were muscular gemsbok with their black-and-white faces and rapierlike horns, slender and graceful springbok antelope, and kudus, large browsers with huge spiral-shaped horns. It was early. This was only our first day in the field.

With each bump, the load on the roof rack shifted, pinging and scraping against metal and wood. The bulky petrol cans screeched forward. Was the cable giving? Was it the rack itself? Stopping to look was an option we couldn't afford. The bakkie in front of us continued its relentless pace, churning up billows of dust.

"Faster, follow the dirt plumes . . . hurry, we're losing them. I can't see!"

Mopane trees gave way to shrubs. House-sized euphorbia plants appeared, erect and soft green. Distant buttes shimmered pinkly in the fading light. Then we entered a canyon, and suddenly the air seemed to blow from a furnace, blasting us with torrid winds. Sand appeared. We plowed across dry rivers.

"Maybe I should go into four-wheel low?" I was beginning to worry whether our tires could continue to churn through the thick sand.

"No, just keep going," Carol said. "We've got to keep up with them."

"I'm trying, but we're sinking."

Beads of sweat converged into rivulets down my back. The canyons grew deeper. I imagined hyraxes, baboons, even leopards.

Above Carol's head, I noticed the metal rim which supported the roof rack had now caved down above her door.

She responded distractedly, "It's okay, just keep driving. . . ."

A fire burned at the side of our path—a campfire. Lean dark bodies were gathered around it. In the dim light we could make out shirtless men in tattered shorts, standing, talking, and laughing. The government bakkie drove past them and stopped a few hundred meters up the dusty track. I cut the Land Rover's engine; the momentum carried us silently forward. Jovial chatter stopped abruptly as the men turned silently to face us. Not sure what to do, I opened the door, whisking Sonja out of her seat. Hostile stares came our way. Sonja radiated heat in my arms. Carol remained in the Land Rover as I moved slowly toward the men.

"Goeienannd, hoe gaan dit?" someone said.

Not understanding, I continued my approach. A person moved to block my path. Except for the "United States of America" I understood nothing. The men spoke local languages or Afrikaans. I lifted Sonja above my head and waved her hand. The men looked. Slowly, smiles came.

Darting flames silhouetted women and children against huts. Extended families. African societies view kids differently from most of the westernized world. Life together, communal rearing, is much more common. I hadn't planned to diffuse tension by bringing Sonja over, but the people seemed to now be more accepting of us. The poor kid had been strapped in her safety seat for 12 sweltering hours. It was just a father and daughter in need of fresh desert air.

We drove on to join Pete and Louis and the others, and found them setting up their camp.

The compressed part of the Land Rover's roof rack had pinned itself onto Carol's door; it wouldn't open. Nonplussed, she climbed out through the window. Sweat and dust cemented her hair to her scalp. She greeted the others.

Because our supplies had been completely rearranged during the chase, we couldn't find our flashlight or tools and had to borrow some to saw bolts from the rack so that the door would open. Even the petrol gauge was giving us trouble. It failed to register any fuel, but this couldn't be true. We had just filled the tanks in Khorixas. At least we were here, at the dehorning operation. The American hillbillies had arrived.

Perhaps skeptics of this project had been correct all along. During our earlier planning and correspondence, Raoul du Toit, a Zimbabwean monitoring black rhinos in his country, had written, "What we don't need are more Americans bumbling around in the bush."

It was nearly 9 P.M. Dinner had been on our minds for hours, but this was our first chance. The bread and fresh fruit from Windhoek's supermarket had been pulverized during the ride. Uncooked canned wieners seemed safe. Sonja ate them willingly enough.

As we tried to sleep, I couldn't stop thinking about what might be ahead. The heat where we were, in Guantagab Canyon, was oppressive, and we weren't even in the real desert yet. I was beginning to understand why studies like those done in the States or Europe weren't done here: logistics. Distances. Communications. Questions nagged. Could we find rhinos? How dangerous were spotted hyenas? We knew that lions in the Kaokoveld had been persecuted and were afraid of humans, but we planned also to work in Etosha National Park, where we had heard lions were abundant and aggressive. A barking gecko interrupted my worries, and I fell asleep listening to a distant jackal.

By 6 A.M. the air had cooled to 80°F. *Poor Sonja, poor Carol. Did I really drag them to Africa?* Doves cooed. Morning would soon be upon us.

3

Trial by Fire

[Carol]

Between bouts of sleep we gazed at mud-colored canvas—our new tent. Mosquitoes bounced against the netted windows like lions anticipating a Christian feast. Other team members used cots strategically placed below finely meshed nets dangling from branches. This African approach hadn't occurred to us.

I wondered how Sonja would tolerate long days of searching for rhinos. It was bad enough bringing a baby without having told anyone in advance, but, as outsiders, it would be worse if we held others back. I hated to think of another possibility—that I'd be relegated to camp because of our decision to have a child.

At least Blythe Loutit was coming. An artist by training, she lived in Khorixas and had savvy, having lived in the deserts for 20 years. As field director for Save the Rhino Trust, she knew about rhinos and elephants. She'd won the Sir Peter Scott Award in England for her untiring efforts on behalf of conservation, and she had encouraged our project. The prospect of meeting and working with her was exciting.

We woke before dawn that morning, our first in the African bush, to enjoy coffee and just savor being there. A small bright light floated across the eastern horizon, startling us.

"Hey, Cunningham," Joel said, "this is ridiculous, but we don't believe in UFOs, do we? So what is that? It can't be a satellite, can it?"

I had no answer. Later we asked others, but no one had seen anything unusual.

For three days we remained bewildered until a report on the shortwave radio restored our sanity. The space shuttle had been visible in the southern hemisphere.

Ten rhinos needed to have their horns removed. Blythe believed they were too close to humans and in danger. Her recommendations were valued by the ministry because of her dedication and long experience. Within the animals' range was Guantagab, a poor mining community full of shacks, squalor, and broken glass. A baby rhino had been stoned to death near the settlement four months earlier and the hungry Damaras had eaten part of it. Blythe feared that soon the money that adult horns could bring would be too powerful a temptation for these poor people to resist.

Since the 1970s, black rhino populations have plummeted throughout Africa. The northern Namib Desert has been no exception. Namibia, the driest country in subsaharan Africa, is sparsely populated. It is also large. With less than one and a half million people (in 1991) and 823,144 square kilometers in area, it is twice the size of California, 50 percent larger than France.

Namibia's population density is the fourth lowest in the world—1.6 people for each square kilometer or 4.2 people per square mile; higher than Greenland but lower than neighboring Botswana. In contrast, Kenya has 44 people per square kilometer, or 115 people per square mile.

The first Namibian rhinos that had been dehorned lived north of Doros Crater. No one said much to us about that operation two years earlier, but we were told that all of the rhinos were doing fine, two births had already occurred, and dangerous predators such as lions and spotted hyenas were rare in these regions. By dehorning the animals and assuming that they were then safe from poachers, scouts of Save the Rhino Trust could concentrate their activities in other areas. The logic made perfect sense, but occasionally, when we'd ask questions, people became edgy. For the time being we decided to curb our curiosity. A rhino without a horn was certainly better in the eyes of its Namibian protectors than a dead rhino. It was better off in our eyes too.

The words of Eugene Joubert haunted us: "You'll be doing well just to find one rhino each week." Now his warning made sense. The area inhabited by the rhinos of the northern Namib Desert is huge, perhaps 30,000 kilometers square. It is an open country full of fractured mountains, torturous canyons, and reed-filled flood plains. Gravel and brutal rock stretch for miles, shredding anything less tough than 10-ply tires. According to Blythe, a single rhino might roam more than 1,000 square kilometers, an area larger than many national parks.

No wonder the dehorning team was having poor luck finding rhinos. Even with a helicopter, 8–15 seasoned trackers, half a dozen vehicles, and people who had worked these deserts all their lives, the work was difficult. We had no idea how we were going to do this study, the two of us and a baby.

On the third day we caught up with a pod of other vehicles, just in time to see the helicopter leaving. A darted rhino lay on the ground, one known as *Sandy*, who Blythe had named 10 years earlier. Two men dragged a two-handled lumberman's blade back and forth across his horn. As in a carpenter's shop, chips and dust from the horn littered the ground. With Soni tucked snugly into the babypack, I scribbled notes furiously. Joel measured the rhino's head, its body's girth and length, and its feet, and examined its massive jaws. Men from the team doused the drugged animal to keep it from overheating.

As the dehorning operation spilled into its second week, we ricocheted, sweated, and smelled. Worse, there was still little success in finding additional rhinos. We consistently delayed the Game Capture Unit. We witnessed exploding tires, engine failures, and overheated government vehicles, and the incentive for not assassinating our Land Rover was strong. But others grew impatient with our slower pace, and when we ran out of petrol—though our two fuel gauges read half-full—the lingering caravan became testy. Any repairs meant a stay in the capital of a week or longer. The Land Rover was our home, and we couldn't afford to be without it. During this operation, adding more petrol and worrying about gauges later would have to do.

On the eighth day, we decided to make a supply run to Khorixas. We asked Blythe for directions to get out of the puzzling maze at Doros Crater. "It's complicated," was all she said, smiling.

Her attitude puzzled us but we needed to find our way to town. For the next three hours we angled in what we thought was the right direction. Finally we stumbled upon an old farm, hardly more than a shell. A lone Damara stepped out. When we said "Khorixas?" he simply pointed up the dirt road we'd been following.

In town, our experience soured. Joel gagged on his first gulp of fresh milk. The carton was not dated, the milk not pasteurized. We were learning about life in small African towns. At the only service station, the air pump wouldn't pump air.

From our journal. Day 9, April 10. Another day of rhino-chasing and chaos. We drove 140 kilometers, last as always. Two German photographers blew a tire during one high–speed pursuit. We found their car abandoned and figured that the chopper had swooped down to get them. Apparently all the whites piled into the chopper, to get to the rhino, and the trackers walked off to find their leader, who'd been left behind (10 kilometers away). . . . We hiked several ridgetops hoping to see the sun glinting off one of the metallic vehicles. Nothing. We discovered we were alone in the hills and returned to camp. They had dehorned another rhino and we'd missed it.

As the days passed our tactics changed. The Land Rover now stayed at camp with Soni and me. Because we'd been slowing the operation, JB (Joel) rode with the trackers in the open bakkie. The temperature was up to 104°F. While he bounced through hour after hour and shared cans of grapefruit with the men, we made the best of life in the canyon. We bathed in plastic wash basins, I with four wonderful liters, Sonja with six. She loved her first chance to be cool.

We also had annoying visitors—mopane flies, actually tiny bees, which buzzed, crawled, and nagged. Sonja couldn't nap. I loaded her in the babypack, and together we walked. The flies were less irritating as we moved, so she slept rather than cried, and I got exercise. Much later, when the incoming helicopter's whock-whock-whock shattered the silence, I raced to it to learn if the "men" had had success.

One afternoon after the operation quit early and the trackers received their first rest, Blythe tantalized us with a brief look at her photo albums of rhinos. Prior sightings with names and locations were recorded. She let us glance through them for a few minutes, then she closed the book. She pledged access later, a promise that excited us because we had to know about the ecology and behavior of individual rhinos before the dehorning if we ever hoped to understand what happens after horns are removed.

Another day, Blythe suggested that one of her trackers ride with us. A lean, quiet man with muscular legs, he spoke some English and was well acquainted with these deserts, having grown up in Sesfontein. As we drove, I noticed a tattoo on his left arm, a conspicuous dagger. We had heard of him—he was a convicted felon. While working for Save the Rhino Trust, he had poached a rhino. His Damara name was !Huie, but he was known to Blythe and to us as Archie, Archie Gawuseb. He seemed nice enough, and we hoped he could teach us about rhinos.

More rhinos were darted and dehorned, but often we weren't there and lost valuable data. When a visiting veterinarian tossed tissue from an ear notch into the dirt, it rotted under the powerful sun. Somehow it hadn't occurred to him that it was needed for our study. The language barrier mattered. And, when we searched for it the next day, it was gone.

Our study was aimed at understanding possible relationships between horns and reproduction. One goal was to determine which males sired most of the offspring. To learn if dehorning had biological consequences, we wanted to know whether mating status had changed for males who lost their horns. Perhaps dehorned individuals would be subordinate. Tissue samples were critical, because DNA tests could reveal who the most prolific fathers were. But without samples, we couldn't address these issues, nor could we provide the government with information about the biological effects of dehorning.

Finally, after nearly two weeks, the eighth rhino was immobilized. Joel had been present at four dehornings, I only at two. When the team returned each day, JB quizzed its exhausted members about ear notches so that we could identify the rhinos as individuals later.

At camp the next morning, the collected horns were displayed. Some long and thin, some thick and smooth, they stood erect on a small camp table to be photographed and examined. Blythe's husband, Rudi Loutit, the chief nature conservation officer of Damaraland, took measurements. Joel asked if he too could measure them. It was an uneasy situation. Joel was interested in additional data, and he also knew that variation can result when different people take measurements. He also didn't know how to ask without offending Rudi. Finally, when Joel mustered the courage and asked as innocently as possible, Rudi's face reddened and he raised his voice and barked, "You can get all the measures, next week, at my office in Khorixas."

The dehorning team left. Abruptly, we were on our own—usually the way we liked it. But we still hadn't gathered any data, and now we sat in a remote corner of Africa with Joel ill, unable to keep food down. He kept saying that he'd be fine, but then drifted into sleep, mouth open, sweat dripping. He slept nearly 16 hours a day for three days. I worried about malaria and tick-borne diseases. Fortunately, it passed. Weaker, but with typical energy, he rebounded.

Before returning to Etosha, we tried to find rhinos on our own. Searching for a fountain (a spring or water hole) near a cave, we followed an unnamed riverbed that cut between canyon walls somewhere to the southwest of Doros Crater. Mountain zebras and rhinos had come this way, their hardened dung indicating that they had not been there recently. Other signs signaled more life. Vertical streaks of white glowed on the rocks in cone-shaped drips, evidence of rock hyraxes and dassie rats. We ignored the shade of a few leadwood trees and kept going. It was 10 A.M., and the early sun was already ruthless. Finally, under the outstretched arms of a tall camelthorn acacia, we rested.

We continued hiking along a zebra trail maintained by centuries of hooves. Sonja sat strapped into her pack on Joel's shoulders. Boots crunched on fine stones as they talked quietly. Five kilometers later we saw a huge vertical rock above an alkaline water basin. Cut into the rock was a cave. Rhino dung littered the ground, several heaps nearly half a meter high and others four meters long.

Dung has unusual significance for many mammals. In species like horses, zebras, and rhinos, it plays a central role in olfactory communication. Males often deposit feces in conspicuous areas and then spend enormous amounts of time sniffing the odor-filled nuggets, much more than females do. Besides the obvious function of ridding the body of undigested material, feces telegraph information that might include individual identities, social status, reproductive state, and body condition. The dung of some species advertises territorial boundaries. The specific significance of rhino dung piles, however, remains a well-kept secret.

Because rhinos are dangerous and are known to kill people as well as lions and elephants, we cautiously checked whether they might be resting in the reeds that grew tall, almost forestlike, around the water. Deliberately, we talked loudly, craned our necks, and listened, both of us too afraid to enter the cool, shady thickets. From a safe distance, Joel lofted handfuls of tiny pebbles into the air, hoping the downward spray would warn any rhinos of our presence. Then, we traded off staying with Sonja and took turns inching ourselves up the talus slope, rock by sizzling rock, to reach the cave.

Cool refreshing air waited. Intricate Bushman paintings lined the walls, the elegant work of past hunter-gatherers. Human and animal figures stared back at us in ocher. Rock cairns, bleached by centuries of sun, had been left by the Khoi, the Bushmen's distant ancestors.

Just before we left, JB noticed some bones piled up against a rock wall. A charred odor drew us closer. Burnt skin lay nearby. Someone must have enjoyed a feast. The skull was missing, but the thick, graviportal bones made us think it must have been a young rhino. We placed a few chunks of seared flesh in a plastic bag so that it could be analyzed later and started back up the sizzling canyon.

At the last shady alcove the tiny thermometer hanging from Joel's pack read 107°F. Sand sucked on our boots, and the lack of trees hadn't stopped determined mopane flies from assaulting us in their search for moisture. Sonja had been crying for 20 minutes. Sweat dripped as we walked like automatons, hoping to find another shade tree. The trail disappeared. I was beginning to hate this project, even Africa. We hadn't seen a single rhino on our own. Joel had lost weight, and now we baked in this bleak canyon. Sonja's tears turned to screams trumpeting into Joel's ear.

"So, are you glad we came to Africa?" I asked, feeling like a bitch. JB walked faster, distancing himself.

4

Etosha

[Carol]

Returning to Etosha National Park, we sailed on blissfully smooth tarmac, heads bobbing to the tunes of Bruce Springstein, Neil Young, and Raffi. As the 300 kilometers that separated us from the park closed rapidly, the mopane woodlands grew thick again. Squat bushes hid water troughs and miles and miles of fences. The country had lost its wild flavor. This was "farmland." Stringy cattle poked from shady refuges. Hands waved from donkey carts. Bright yellow signs warned of straying kudus.

* * *

"You have one of the most difficult projects in Namibia," Dr. Malan Lindeque, the new director of research at Etosha's Ecological Institute, told us, when we met with him after we arrived. He was young, exuding confidence and energy as he explained the challenges ahead for us. We'd have to be prepared to live like nomads, not only in the desert but even at Etosha. Aerial support would be non-existent, the distances huge.

Namibian rhinos had survived, now second only to those of South Africa in population size, not because of sophisticated planning or protection but by default: Geography, low human densities, and remoteness had played crucial roles. To stem the rhino killings that were sweeping across the continent, Namibia had pinned its hopes on dehorning and translocation. (In Kenya, dehorning had been vigorously debated as a conservation strategy in the early 1980s but was never tried. Instead, their remnant populations were sequestered in small, well-guarded sanctuaries. Zimbabwe maintained a stronghold of about 2,000. The status of rhinos in nearby strife-torn Angola and Mozambique remained unknown.)

We told Malan that we hoped to help the rhino conservation effort by gathering information on the efficacy of dehorning, on reproductive rates, sex ratios, dispersal, and behavior. Malan explained that the dehorning issue was polarized in Africa. Because of regional conflicts and differing views on the elephant ivory trade, hard feelings endured, and neither eastern nor southern African biologists were welcome to do such a study. A neutral party made sense: us.

Namibia's Rhino Advisory Committee would soon convene. Malan invited us to the Ecological Institute for orientation and to discuss how our studies could contribute to their programs.

The Etosha Ecological Institute is in Okaukuejo, the westernmost site of public tourist camps in the park. Its purpose is to carry out scientific research and implement effective management for Etosha's wildlife. In 1876, Gerald McKiernan, the first American to describe this part of the world, wrote: "Okoquea is on the edge of the Ovampoland plains, a permanent fountain and much frequented by game of nearly all kinds found in the country." He found "great numbers of wild animals feeding on the open plain. Gnus, zebras, gemsboks, hartebeeste and thousands of springbok were before us, and above the low bush to our left were the long necks of six giraffes. It was the Africa that I had read of in books of travel. All the menageries in the world turned loose would not compare to the sight I saw that day."

Okaukuejo is home to the administrative, research, and management personnel of the Institute and the park, and other workers and their families. The floodlit water hole lures thirsty animals, much to the pleasure of the 100,000 people who visit the park each year. Ironclad gates fitted into a large stone wall swing open at sunrise and close the compound at sunset. No tourists travel the park's wild areas after nightfall, nor may they leave their cars when outside the rest camps. Beige stone buildings line Okaukuejo's roadway, but the most notable feature is the old, tall "German tower," dotted with staggered rectangular openings.

We were nervous. The heavy hitters of Namibian conservation would be at this conference. Hanno Rumpf, the Permanent Secretary of the Ministry of Wildlife, Conservation, and Tourism, would introduce us. Deputy Minister Ben Ulenga would be there, as would Eugene Joubert, Malan Lindeque, Blythe Loutit representing the Save the Rhino Trust, and other management and research officers from around the country.

Sonja stayed with Malan's wife, Pauline, a biologist, and their young son. As we pulled away from their home, loud sobs blared above the engine's hum. Sonja had spent barely a minute away from us since we'd come to Africa. Joel and I looked at each other. The cries faded as he accelerated. She'd just have to endure. We couldn't miss this meeting.

Among the seasoned field vehicles that blocked the walkway to the Ecological Institute was a new white BMW with green government plates. The top administrators had arrived. Knee socks, olive-green shorts, and epaulets filled the halls. Only Hanno Rumpt and Ben Ulenga were not uniformed. We were obviously the Americans. Fortunately, our good clothes weren't very wrinkled, and Joel's hair was short.

After formal introductions came formal apologies. Malan apologized for missing the last meeting: "I regret being out of the country."

Ten long solemn minutes passed.

"Mr. Chairman," said Eugene Joubert finally, "there is a long list of items on the agenda."

He went on to suggest that, rather than each person addressing the chairman for permission to speak, these formalities should be dropped. Rumpf disagreed, addressing Ulenga as "Comrade Chairman." Joel shot me a look of frustration. This meeting would be long. Poor Sonja.

Numerous topics followed, some boring, all relevant: President Nujoma's request to harvest local game for starving people, the translocation of rhinos out of areas near farmers, issues associated with the general treasury, and why funds derived from conservation could not be plowed back into programs for conservation. Everyone agreed that the solution was to find a way to cycle money to local people to reenforce the value of protecting wildlife.

Finally, it was our turn to speak. I highlighted our project's goals, indicating that we wanted to evaluate the effectiveness of dehorning. Our plan was to study horned and hornless rhinos in the desert and also to use Etosha's animals because they showed natural variation in horn size. Malan's hand darted into the air. He exploded, saying that our initial proposal had been rejected because it was unacceptable to concentrate on horns alone. The room fell silent as all eyes focused on us. We didn't have the slightest idea what was going on. Malan's boss, Eugene Joubert, had supported our project from the beginning. Our faces reddened as we tried to think of something to say. Malan had even written us that our proposal was just the type of research needed. Now his outburst muzzled the room.

Joubert intervened, indicating that we were in Namibia to work with the dehorned animals. Joel suggested that these animals needed to be compared with horned animals in the desert and, in the park, of course, information on population dynamics, individual reproduction, and behavior would all be gathered. Malan seemed pleased and suggested that we get together with him to iron out details. We breathed easier.

Once the meeting ended, we sought Malan. We were confused about what he thought we were doing in Namibia. Our project had been approved by the ministry's head office in Windhoek, and an official copy Malan certainly had. His prior correspondence had lauded our aims. In his office, now jovial and expansive, Malan explained that he really didn't care what we studied; all he considered important was that we compare the horned animals from Etosha and dehorned desert animals. As far as he was concerned, he said, "the most important data to us are population dynamics; horns are secondary." We nodded agreeably but were confused by his unpredictable behavior.

We stayed temporarily in a small bungalow in the tourist section of Okaukuejo's fenced compound. There wasn't much space; for exercise, some people ran laps or climbed the stairs of the old German tower. We sought the two Americans we knew from Windhoek, Bill and Kathy Gasaway, who had settled in Okaukuejo, hoping to study Etosha's predator-prey system.

"Jogging" was Bill's remedy for camp lethargy.

"Where?"

"At the airstrip just outside Okaukuejo. We'll show you." He seemed overjoyed to share this small pleasure.

Hardly any of the Etosha staff jogged yet planes were rare, the terrain was level, and vegetation was low.

Carol and I eyed each other, knowing that the Gasaways had already noticed our guarded expressions.

"It's reasonably safe," Bill said. Kathy nodded and smiled to reassure us.

Two hours later we were at the airstrip by ourselves, sitting atop the Land Rover and scanning for lions. It didn't matter that most shrubs were less than half a meter tall, widely spaced, and unlikely to conceal lions, or that the barren runway extended nearly a kilometer in each direction, or that it was still 87°F and lions were generally inactive in heat. We were going to be careful. Slowly we drove up and down the runway, stopping often to examine footprints for large kidney-shaped inverted pads, four toes, no nails.

"Do you think this is lion?"

I didn't think so and shrugged.

Joel jogged first. I watched him shrink as he moved further and further from the safety of the car. Springbok and plains zebras stopped grazing to stare at him. Then my turn came while Joel entertained Sonja. She waited, not too patiently, as he alternated between sentry and paternal duties. The exercise was enjoyable, but our fear of lions remained.

During the next week we explored Okaukuejo and the park. At the Institute, the management staff agreed to let me use the darkroom. We'd provide chemicals and our own paper. Also, Joel could have a small desk during the short periods we'd be in Okaukuejo. Telephones were more of a problem. All calls had to be placed during work hours from the post office.

At night we enjoyed Okaukuejo's floodlit waterhole. We continued airport jogs, and thoughts of lions made our cardiovascular workouts that much better. We wondered if we'd ever be comfortable jogging there.

5

Dark Nights and Moonlight

[Joel]

Fifty wildebeest, large and black, crowded at the water's edge, calves bleating anxiously from somewhere within the jostling mass. We watched from the top of the Land Rover and nibbled a dinner of canned fruit, peanut butter, and Pro-Vita crackers. Our preliminary work in Etosha National Park was beginning. We were at a water hole, checking how well our night vision equipment worked without much moonlight and whether many rhinos drank there. The name of the water hole was a mouthful, Ozonjuitji M'Bari, M'Bari for short.

A small group of zebras entered the clearing. Sonja began chortling "zee, zee!"—each time louder. In unison the animals bunched and ran, with half a dozen ostriches lurching around them. Panicked eyes and hooves disappeared in choking dust. Wildebeests scattered. Amazed by the commotion, we looked at each other.

"I think it's bedtime for Sonja," Carol said. "I'll take her down and read to her for a while. See you later."

Unlike desert rhinos, which were rare and so widely spaced that we'd be forced to track them during the day, Etosha's rhinos could best be seen by waiting for them to drink at night. This was the first of several water holes that we'd check in the huge 22,700-square-kilometer park.

We hoped to find three areas where we could identify rhinos and watch them interact. Park staff had suggested several water holes that had good rhino use, but we also wanted sites with extensive open areas around the water. Otherwise, rhinos could easily disappear into the thick thornveld and we'd lose valuable data.

This requirement had another, more primal origin—fear. Tall grasses could hide lions. If trees were abundant, elephants could arrive unannounced. The safest places would be large vleis (pronounced flays), nonwooded sites with little or low vegetation near the water to guarantee good visibility.

Inside the Land Rover, Sonja finally slept, her fragile body tucked into the back seat. Netting draped the open windows and frustrated buzzing mosquitoes. It was 8 P.M. Carol joined me as I assembled the night vision equipment and tested it under the already star-studded sky. Jumbled waves pulsed on a green screen driven by two AA batteries. According to the brochure, we would be able to see with starlight, which would be magnified 30,000 times. We hoped to detect rhinos in the darkness from 100 meters away. But with only a sliver of moon to help, we wondered how well this would work. Waiting for rhinos to come, we sat in the stillness, dwarfed by the luminous sky.

The sounds and sights of technology were absent—no flashing lights, no vapor trails, no engine hums to distract us from the night and its creatures. Our ears tuned to the chitter of bats and the swoosh of a Pearl's spotted owl as it dived.

We also heard unnatural sounds broadcasting our uninvited presence. The mattress we were sitting on breathed and sighed. Our shorts scraped. When we climbed up and down the ladder the car rocked and squeaked. Our three-layered Gore-tex jackets, wonderful high-tech clothing equipped with zippered underarm vents, double storm flats, and velcro pockets, roared alien sounds, impossible to silence. We tried not to move. The hours ticked by. No rhinos.

"What's that? Give me the scope," whispered an excited Cunningham.

We strained to see our visitors. But without a decent moon, our night vision equipment was as useless as our binoculars. Magnified starlight was greatly overrated, at least in this situation.

"Wheee, wheee, grrr, grrr" came from two dark shapes. Seconds passed.

"Pff, pff."

Unable to tell what was out there wheezing and blowing, I aimed the spotlight, 750,000 candlewatts of white light, at two immobile contours. Diamond-shaped eyes glowed. A springbok and a wildebeest tensed for blast-off as I switched off the light.

"Let them be. They must be thirsty," Carol sighed.

Our inexperience showed. That Zimbabwean, Du Toit, was right. We were "bumbling in the bush."

Unlike rhinos in the Namib Desert, Etosha's rhinos were marginally social, at least according to local dogma. They drank in darkness, synchronously, as many as 12 appearing at a water hole. Similar behavior had been reported 50 years earlier from Kenya. Two filmmakers, Martin and Osa Johnson, had written of their African experiences and marveled at the unusual phenomenon.

We continued our vigil on the gloomy vlei, trying to squelch our feverish imaginations as we sat. Gray termite mounds became elephants. Each thicket concealed a leopard. We thought of Swedish explorer Peter Moller, who, after visiting northern Namibia in 1895–96, reported: "The natives traveling through this area . . . build structures in trees where they spend the night to protect themselves from the lions." Now we worried that our roof rack wasn't high enough to discourage inquisitive lions, and knew it was too low if elephants charged.

Like many primates, we humans are ill-adapted for the night. Baboons seek the safety of trees or cliffs after sunset. Our human ancestors used caves. Work-driven, we sat exposed and struggled to stay awake. Carol nestled against my shoulder. I reciprocated, leaning against her. Our heads nodded as fatigue closed in. Our jackets still rustled. It was 1:05 A.M.

My eyes opened. I breathed deeply and felt my heart pumping adrenaline through my body.

"Look, at the water," I whispered, jabbing Cunningham lightly.

Two tanklike shapes were nearby.

A short "Plghhh hhhhhHHHGGHH ooooough" fractured the quiet. These couldn't be wildebeests or springbok.

One hulk whirled and faced us. We sat straighter, our hearts still pounding.

"Should I flash?" I wondered aloud.

"Okay."

I engaged the spotlight. Eyes glowed from horned faces. Abruptly, the shapes became rhinos accelerating away into darkness, the ground shuddering with their weight.

"Next time, try the red filter," Carol muttered in disgust at our incompetence. It was 1:18 A.M. Maybe they'd come back.

At 3 A.M., two more rhinos appeared in the gloom. They might have been the same as the earlier pair. We didn't know. We had been too excited to notice the sexes of the earlier visitors. These two were probably a mother and her calf. Adult males and females are the same size, but these two were not. They stood still for several minutes before retreating into the woods.

For our study we needed photos at close range. Otherwise it would be impossible to identify individuals and recognize them later. Horn shapes, ear tears, even lip wrinkles would provide the key. John Goddard had used drawings to identify East Africa's rhinos more than 30 years earlier, but it now looked as if we'd be lucky to see any. We questioned the $7,000 we had paid for our night vision equipment. If we couldn't see rhinos from atop the Land Rover, I certainly wasn't about to approach them on foot for photos. Like Carol, I too was wary of rhinos and lions, and the dark.

At 4:30 A.M. we aimed the spotlight, now covered by a red filter, toward another pair, large and small, definitely a mother and calf. Again we saw the runaway train response—chuffing and heavy breathing followed by pounding feet—but this time with a violent charge at the Land Rover. With startling agility and speed, they veered off at the last minute, vanishing into the night.

On the third morning at M'Bari, daylight came softly.

"Mommm, Mom-my," Sonja called from below. Delicate springbok grunted and fed 100 meters from our motionless, lumpy forms buried beneath a blanket. As we raised the blanket's edge to look out, the first rays of sun touched the velvety horns of a giraffe, its head pitched forward against the sky.

Despite our delight in the veld's beauty, we returned to Okaukuejo, low in spirits and exhausted by our three frustrating nights.

A new strategy was needed. We could always watch at the local floodlit water hole. Visibility was excellent, and the rhinos were already habituated to human shapes, raucous voices, and laughter, even the smells of meat braiing on outdoor grills. The Okaukuejo water hole was not the Africa we'd imagined, but it was an option we couldn't dismiss. We needed data. We also needed to talk to someone, someone with knowledge who would be willing to talk with us.

Blythe Loutit had tracked rhinos during the day, but knew little about behavior or biological work. Eugene Joubert had studied both Etosha and desert rhinos, but not at night. Malan Lindeque had worked mainly with elephants, and the chief warden had the park to manage.

We sought advice again from Bill and Kathy Gasaway. The two ex-Alaskans had worked in Etosha on and off since 1986 and knew many of the rangers and administrators. They suggested we meet with an experienced ex-hunter who had spent 20 years in the Congo, Raymond Dujardin, of Belgian descent, who now served as the head ranger for Etosha's central district. He lived some 70 kilometers away at Halali, another tourist camp within Etosha. He was usually happy to share his experience with newcomers.

We repacked the Land Rover and headed to Halali. Bill and Kathy met us there to make the introductions.

Raymond's gentle smile and warm manner immediately put us at ease. He was curious about our study, our objectives, and why we wanted to learn about rhinos. Not once did he say it couldn't be done, and, to our delight, he took an interest in Soni, talking and playing with her.

"Ooo-doo, ooo-doo," she said, pointing to the horns on the wall behind Raymond. When I explained that she meant "kudu," she was his friend; he was impressed that although she could barely talk she already identified animals by the horn shapes alone.

Raymond described rhinos, night work, and precautions. He told us that Africa was *very very* different from Europe, that we needed to be careful. Then he surprised us by saying that he'd take us out for a couple of evenings and teach us some simple bush skills.

"Let us meet at 5:30 P.M., near the spot where the dead elephant is at the water hole. Do you know the place, Joel and Carol?" Raymond asked. "If the lions are not there, the rhinos will come. Joel and Carol, we must be careful and make sure there are no lions. Let us hope that the clouds are not too thick so that there is enough light from the moon."

During the day we tried to get Soni to sleep so that we could rest, but it was hopeless. Neonate naps didn't mix well with flies and heat. Instead, we played with her, and talked about the upcoming night. A recent publication on rhino monitoring by Allan Cilliers did little to allay our fears:

> The observer patiently waits *on the ground* [our italics]. . . . The assistant must be alert at all times. . . . This person must keep a constant surveillance of the surrounding area for other rhinoceros and particularly lion and elephants that may present problems. If likelihood of trouble, either when the rhinoceros that is being photographed charges, or lions or elephants are in close proximity of the waterhole, the spotlight without the red filter must be used to dazzle them and frighten them off.

It had been a long time since we, as humans, had been conscious of our place in the food chain, something our ancestors joined regularly. It was clear why paleolithic peoples were easy meals for cave bears and sabered cats. They lacked effective protection. During our nocturnal vigils the pressure on Carol would be great. A single mistake would have grave consequences, more than any errors I might make during my short wanderings in the dark. If I didn't get a photograph, there would be later opportunities. A lapse in vigilance or judgment by Cunningham would be disastrous for the person on the ground, me.

We met Raymond 100 meters from the waterhole. A .44 magnum handgun was strapped to his hip. He began our lesson by reviewing the night's protocol.

"We must be very quiet. Rhinos usually come from the other side of the water. Joel, you must follow me closely as we approach the rhino. Don't worry. I will watch for other rhinos."

He'd also be the lookout for lions. I should just concentrate on photographing the rhino.

The photographs were important to confirm individual identities, which in turn was crucial in estimating population sizes, information important for the park's monitoring program. During a night or census period, if the same individual came to drink half a dozen times, and if we couldn't identify individuals, we might count it as six different rhinos and inflate our estimates of how many lived in the area. If we recognized individuals, we'd know with certainty that there had been only one visitor.

Shy rhinos, like kudus, gorillas, even grizzly bears, and other species that live in thick bush, are not easily counted. Aerial work is expensive and can miss animals resting in the shade. Photo identities were fundamental, a critical first step for a successful project.

"Do you know much about guns?" asked Raymond in his heavily-accented voice.

My mind flashed, *Boys will be boys.* Language barriers mattered little.

"I'll show you mine if you show me yours," I said, smiling.

Raymond examined my .44, and I feigned enthusiasm for his. Both were big, shiny, and dangerous. Carol and I had practiced shooting before leaving the States, but we had no great affection for guns. Their technical appeal was overshadowed by headlines reporting too many tragic accidents and incidents of mindless violence.

We visited, waiting for darkness and our crash course in ground photography. The procedures sounded simple, deceptively so. There must be a light source on the subject. Focus the camera and shoot. But at night, and at only 20–30 meters from a rhino, the zone between safety and disaster contracts, leaving no margin for error. Rhinos can gallop at 50 kilometers per hour. I wasn't happy that I needed to be so close. And there were logistical problems.

First, we needed to light up the animal while not frightening it away, but most flash units won't illuminate distant areas. Shooting at night is usually for the professionals of *National Geographic* and the film industry. They simply use powerful lighting units. But technology is one thing, shy animals another, as we had learned at M'Bari only a few days earlier. Bright spotlights seemed most effective at lighting up the behinds of terrorized rhinos. What we needed was a fast lens and a flash unit that could light our quarry for a mere instant. Light-sensitive film would also increase the distance from which we could shoot.

Second, we had to solve the problem of focusing the lens. In nearly total darkness it's impossible. Somehow I had to know how far off the rhino would be and preset the lens for that distance. Night work would be intense.

"Don't worry. The first four or five times you'll have adrenaline rushes, then it will be easy," Lue Scheepers, a seasoned ministry biologist, had once told us. Remembering this, Carol and I looked at each other with disbelief. How could we walk around photographing rhinos in the dark?

The sun faded behind green hills and left us bunched near our vehicles at the Halali water hole. Shortly afterward, loud crashing sounds emanated from bushes not more than 60 meters away. Fortunately, Sonja had already fallen asleep. Jackals yelped in high-pitched voices as thrashing continued in the thick vegetation. Soon clouds arrived, obscuring the moon. Despite our binoculars and the night vision equipment, we could barely see.

Cackles began, and turned into a shrieking cacophony. Hyenas. They bounded across the clearing close enough to count. Five, then nine. In the dim moonlight, we could make out 17, right in front of us. They dashed across the open arena around the water hole. Soon afterward, we heard bones crunching between powerful jaws; dinner was the dead elephant. Howls and screams sent tingles, wonderful tingles, down my back. Another 8–10 hyenas appeared, zigzagging and running, intent on making their way to the rotting carcass. Our eyes tried to stay with the bouncing dark shapes, yet we repeatedly darted looks at the clouds above, hoping they would part and allow the bright moonlight to return.

Weighing about 65 kilograms, spotted hyenas are larger than wolves and have more powerful jaws, looking like tailless rotweilers with spots. One suddenly emerged from the bushes and stared at us from less than five meters away. We aimed the blazing spotlight at its face, and every feature—facial hair, intense eyes, black nose—seared into our minds. The hyena dashed away.

Raymond and I noticed a bulky grayness approaching, our first rhino. We walked toward it, Raymond with experienced caution, I with anxiety. With each step, dried grass brushed noisily against my trousers. I felt hyper-alive, blood pumping through my veins. The rhino's ears rotated like helicopter blades, twisting and turning with my every step.

The revolving stopped. The ears were now erect and centered directly on us. I concentrated on the long front horn, the "anterior." My shaking hands fumbled with the camera and flash unit. When I peered through the lens, my view was limited. I could barely see the animal.

"Go closer, Joel, closer," urged Raymond, encouraging my suicide.

At 35 meters the flash couldn't illuminate the rhino, and not enough light meant no photo. No identification. No data. We absolutely had to take bright, crisp photos for our project to succeed. Carol stood near the vehicle ready with the spotlight in case more hyenas became interested in us.

Gently, I placed my finger on the shutter release, deliberately moving slowly. Not that it really mattered; the rhino already knew we were there. Through the darkened lens I tried to guess when the profile was sharp.

"Now, Joel, now," Raymond insisted. "Shoot the picture!"

I hesitated, ignoring the frustrated ranger. Seconds passed. I sensed several hyenas approaching but dared not move my eyes from the lens and the rhino's framed horns. Slowly, I pressed the shutter release.

Pffft.

The flash exploded, radiating light into the pitch darkness. Raymond and I were blinded. So was the rhino. I wanted to bolt but still needed another picture. We needed two: a profile for later measurement of horn size, and a facial with erect ears so that tears or notches would be obvious, as well as lip wrinkles or scars. I remembered *the* warning: *Don't run.* I stood my ground.

While our eyes recovered, the rhino turned again, staring in our direction. Precious seconds passed, my life ridiculously dependent on the time it took a pair of AA batteries to rejuvenate themselves for the next flash. *Look at me. Look at me.* I willed the rhino's head into position.

Pffft—light burst from the unit. We heard the rhino whirl in place, but again we could see nothing. We waited, paralyzed, our pupils constricted by the momentary brightness. The rhino stood frozen, confused. Then, as our sight slowly returned, the animal advanced toward the water. We retreated, my body drenched in sweat, hands still shaking.

The sideshow continued as 30–35 hyenas feasted on elephant meat and bone. These loping beasts, strangely unbalanced by their heavy forequarters, whooped and howled. Although we had seen *National Geographic* specials and knew from the works of biologists in Kenya and Tanzania that hyenas were group-living, in this part of Africa large clans were uncommon. Never before had Raymond seen so many hyenas at one time, despite all his years in the Congo, Zaire, South Africa, and Namibia. Our night in the bush with Raymond was just blind, dumb luck. We savored the experience, unable to stop smiling.

The next night there were few clouds and a slightly swollen moon, which lit up the dusty landscape. Again Raymond guided us. This time, our night vision equipment amplified light and magnified images threefold. The improvement was dramatic. Through the lens we saw crested porcupines waddling to the water. Black-backed jackals, tails erect with interest, watched them. Bold springhares bounced like miniature kangaroos; shy kudus, springbok, and rhinos haltingly approached. This night Carol persevered to capture a rhino on film. Now that we had both practiced, we were ready to leave Halili and try on our own.

After thanking Raymond profusely, we chugged westward looking for suitable study areas. The enormity of our project weighed on us despite the success of the last few nights. Etosha was larger than Massachusetts, bigger than all of Israel, more than twice the size of Yellowstone. We still needed to evaluate several water holes as potential sites. We avoided thinking about Damaraland, where we had yet to find a single rhino on our own.

Dozens of solar-powered units pumped precious ground water into concrete basins. Wildlife management was at work. The hands of man had made the priceless liquid available to thirsty animals, opening thousands of kilometers to them in western Etosha's thornveld. We decided an area with three water holes would be our Zebra Pan study site, a fictitious name we used to avoid the possibility of making rhino locations available to poachers.

The three water sources were relatively close to one another, and we wondered whether rhinos consistently drank at one place or whether there was interchange. This was the southwestern edge of Africa's wildebeest range and the only spot left on the continent where plains zebras overlapped with a different species, mountain zebra. This was the eastern edge of mountain zebra country; their numbers increased up and over the mountainous escarpment, out to the west, and in the barren Namib Desert.

Our temporary residence was an abandoned horse camp, a fenced relic of days past when horses were important for patrolling rangers. Soni's stuffed dolls, a teddy bear, a pink rag doll, and wild-eyed Bert, a doll version of a cartoon character, sat among our assorted trunks, pads, boxes, and scientific gear.

Exhausted from night work, we wanted to rest our eyes during the day. But Sonja had thoughts of her own—exploration. Her nappies clung as she teetered on unsure legs and guided herself through the brush. Even in the enclosure, she needed constant surveillance.

Centipedes and millipedes were hidden under rocks, ticks attached themselves to leaves or crawled on the ground, and snakes slipped away, leaving lacework shells of outer skin as deadly reminders. Our mascot became a yellow mongoose that regularly raided our food supplies. During the midday heat, it found coolness by slipping into an underground burrow. We tried something else, flattening ourselves on a cement slab and absorbing whatever coolness remained from the prior night. Ants and flies explored our exposed bodies.

Soni and Carol played with crayons, toys, and books. When it was my turn, Sonja and I used termite mounds as slides and explored purple-podded terminalia bushes, or just gathered rocks. The horse camp offered a charming respite from worries about lions. To the west we could see a string of rugged dolomite hills; the park's western boundary lay on the other side. The time passed quickly.

Whenever we first visited a water hole for evening observations, hours were required to set up. Since our camp was 20 kilometers from the primary water hole we'd chosen to work, we cooked dinners in the afternoon, the fire adding to the already 95°F heat. Sometimes we tossed in potatoes, to eat later during our night-time observations.

In the late afternoon, we followed a dusty red dirt spoor (track) through open woodlands. We knew when we neared the water hole. Red hartebeest and gemsbok exploded in retreat. Ostriches zigzagged in high steps. Springbok pronked, jumping stiff legged with heads held high, like popcorn exploding in all directions. Toward sunset, families of bat-eared foxes searched for dinner. Radarlike ears tilted and turned until they zeroed in on the subterranean motions of ants and termites.

Since rhinos often began arriving just before dusk, it was important that we already be prepared to gather information. Carefully I balanced rocks on elephant femurs or tibias 27 meters from the water, to mark the exact spot needed for a crisp photo. Because rhinos sniff unfamiliar objects, I experimented to see what might attract them. Elephant dung with human scent was adequate, but fresh rhino feces were best for enticing the rhinos to the markers.

The night plan was simple. After Soni was asleep, Carol would sit on top of the Land Rover, scanning for rhinos, lions, elephants, and hyenas. Unlike other researchers who did observations from inside a car, we couldn't. With Sonja sleeping in the back seat, and with clothing and other gear filling the car, visibility was restricted. Plus my safety would be improved if Carol could see 360 degrees from above ground level.

The vlei had good visibility, an area of about 400 by 400 meters, surrounded by tangles of vegetation one to two meters high that spread in all directions. But our confidence vanished with the sun. As we peered tensely into binoculars and the dark abyss, our minds turned bushes and crumbly termite mounds into dangerous shapes. Mostly, it was imagination. There was nothing to report.

On the second day, at dusk, a rhino appeared. A shy rhino, 300 meters distant. Two ears and a massive cracked horn poked above a bush. I handed Sonja from the roof rack down to Carol and heard fragments of *Jack and the Bean Stalk* and *The Jungle Book* for the next twenty minutes. When she rejoined me, the rhino still hadn't moved. I wondered whether it was asleep or whether it was a rhino at all. Carol scanned with the night vision scope.

"Its ears are moving. That's definitely a rhino." A few minutes later she said, "I don't see it any more. It's moved."

Finally the animal emerged from the brush and approached the water. I made my descent. Reaching soft sand, I advanced silently, hoping for trouble-free photos. The first line of rocks indicated that only eight meters remained before I could shoot. My energies were focused forward. Carol tried to become an inanimate object, part of the roof, under the African sky. I had complete confidence that she'd call out or flash the spotlight at approaching danger. I peered through the camera, a high-tech Nikon N8008S, different from the manual one I had used with Raymond. The pulsing of lights flashing inside the viewfinder distracted me. When at last my eyes adjusted, I pressed my finger down. *Pffft*—the flash detonated. Neither I nor the rhino could see. Then *dzzzzzzzz, dzzzzzzzz*, the film noisily advanced automatically. The rhino now knew exactly where I was standing. Waiting, I counted, *one, two, three, four, five* seconds, until, I hoped, the batteries in the flash unit were recharged. I touched the shutter release. Again *pffft, dzzzzzzz*. Two photos. I took a deep breath.

"If all goes well in the darkroom, we'll actually have our first data photos," I whispered to Carol as I stepped up over the roof rack's metal guard rails. Later, after he drank and left the clearing, we discussed a name for him. We settled on *Chama*.

An elephant bull was the next visitor. We noticed him when he was just 60 meters away. The reports were right, we saw that elephants are silent and swift, and taller and heavier than the Land Rover. We held our position as he ambled past us and to the water.

Almost simultaneously, two rhinos arrived from different directions. They acted nervous. We wondered if they were shy because of the elephant, the other rhino, or us, and we didn't know if we should try to get photos now or wait for a more opportune moment. *What if Sonja wakes up and cries?* we wondered. *Would the elephant attack?* Glued to the roof rack, we were back in our role of spectators, questioning the scenes as they developed. It was 7:20 P.M.

First, the two rhinos approached the water, where the elephant bull stood impassively. His tusks were short. As the rhinos drew near him, his ears flared back and forth like huge blankets shaking dust. He looked even more enormous than before. His trunk moved up and away from his body, then it snapped down and forward, missing one rhino, which had turned and run. These rhinos knew that elephants could be deadly.

Early explorers had also observed encounters. In the 1830s James Edward Alexander described interactions he had seen on his travels in Namibia: "When the elephant and the rhinoceros come together and are mutually enraged, the rhinoceros avoiding the blow of the trunk and the thrust of the tusks, dashes at the elephant's belly and rips it up."

Swedish adventurer Charles John Andersson wrote of his 1856 observations in his book *Lake Ngami*:

> strength in the elephant is infinitely superior to the rhinoceros, but the latter on account of his swiftness and sudden movements, is by no means a despi-

cable antagonist. Indeed . . . they have perished together. At Omanbonde . . . a rhinoceros, having encountered an elephant, made a furious dash at him, striking his long sharp horn into the belly of his antagonist with such force as to be unable to extricate himself; and, in his fall, the elephant crushed his assailant to death.

Frederick Thomas Green reported in his 1864 diary:

During our frequent night shootings . . . a bull elephant approached a small hole of water where another was drinking. . . . He passed on and directed his steps to another pool where a black rhinoceros was standing. The latter never attempted to move but, grunting hideously . . . seemed to challenge his adversary to combat as he approached. At length, the elephant observing "Mr Borele" [the rhino] with every appearance of hostile intentions . . . walked off again. . . . Again and again the elephant tried to approach the water, but every time he failed.

Nevertheless, elephants are deadly. A simple swipe of a trunk has killed a 400-kilogram domestic cow. During the 1980s, more than 500 Africans, and an unknown number of tourists, were killed by elephants. We wanted no part of the potential fracas. At 100 meters we were insignificant and unnoticed. There would be ample opportunity to resolve the issue of "interspecific dominance" once our sample of known rhinos increased.

My hands cupped the unwieldy camera with its telephoto lens and flash unit, and binoculars bounced against my chest as I climbed off the Land Rover and moved in. The two rhinos were still too far from my markers, but if they approached to within 27 meters of me, I'd get crisp photos. The focal length of the telephoto lens was fixed with duct tape. I knelt and waited, a small shadowy lump on the ground covered by darkness.

Then, like a desert lizard trying to disappear into sand, I lay on my stomach. The lower I got, the easier it was to detect objects that broke the horizon. My dilated pupils sucked in as much light as possible, and I peered toward the dim gray forms. The details were fuzzy, impossible to make out. One rhino had a large horn. Was it split? Was this Chama? Had he returned with another rhino? I needed to know. If it was him, I'd only need to photograph the other rhino.

I concentrated on the rhinos and waited, unable to move forward because of the elephant. But when he turned in my direction, I hesitated to flash in his. Stalemate.

The rhinos left. No data.

At 2:30 A.M. we switched on the ignition and crawled the 20 kilometers back to our camp. Two crested porcupines disappeared into a thicket along the road's edge. A small catlike animal with thinly spaced spots glowed in the spotlight. It was too small for a leopard, probably not a cat at all, perhaps a genet. Despite bouncing in the back seat because of the rough road, Sonja never woke.

The next day we discovered sabotage at the water hole. All of my markers had been scattered, and the culprits were obvious. The deep, rounded footprints of elephants, both large and small, were everywhere. They had come in the night, dismantling my scented towers with their dexterous trunks.

As we readied for another night, nearby animals were strangely agitated. Zebras bunched and vocalized more than usual. There were only a few springbok. Where were the large numbers? Regularly, there were 150–300 at this time of day. There were no hartebeest, no giraffes. Gemsbok spooked easily.

A huge veil of dust captured the setting sun, sending pillars of golden light in all directions. We enjoyed the beauty, looking forward to the night's surprises, until two gut-wrenching screams shattered the peace. Zebras ran through thick bush only half a kilometer from us. The sun dropped below the horizon as roars sounded.

"Aoohh, aoohh, aoohh," cracked the night. Lions, there was no mistake. And they were close.

"Joel, let's get in the car, now." Urgent. We'd heard these western lions, unlike lions habituated to tourists, were trouble.

Flip Stander, a biologist who had studied the huge felids and who was known locally as the white Bushman, would not jog in this area, even if accompanied by a car. Mickey Riley, a ranger from nearby Otjovassandu, had watched with fascination as a lioness leapt onto his truck's hood and plunged huge canines into the thick metal. Two holes about the diameter of my thumb remained. The Gasaways' Toyota bakkie (pickup) had been charged. Even the fearless rangers who monitored rhinos did so only after they'd used vehicles to chase lions from the water holes.

From inside the Land Rover our view was restricted, as we had expected. It would be difficult to see lions or detect rhinos in advance.

"Maybe they won't come to drink and just stay with the kill. Maybe we should try to see where they are," I said, hinting that we should go back on top, or better, dive through the thick brush looking for them.

"You're crazy. Just leave them alone."

"Why should we sit here in the car? Better to know where the lions are. From in here, we can hardly see. I bet if we get on top, we'll be able to see them."

"I think it's a bad idea, Joel." She said it slowly and clearly, as if I was missing the point.

"C'mon, Cunningham." She is rarely this stubborn. Why don't we just lock the doors? "We might wait here all night. If rhinos come, I want data."

After an hour I was restless and began pestering her again. She looked at me, then started scanning the dark clearing.

"Wait a minute, there's a lion. Damn it, not more than 20 meters away! Four lions! No, five. You didn't even see them." She was disgusted.

With binoculars we could see their muzzles, covered with blood.

"Oh," I said.

Over the next few days we photographed different rhinos. We were certain of at least eight, but there might have been as many as 16 different individuals. Confirmation would come only after Carol developed the negatives and printed the photos at the Institute. We had seen interactions and would know who was doing what to whom; who charged, who retreated, who had calves, and who did not. Retrospectively we would assign names to animals we numbered in our data books. Now we needed the developed prints.

We prepared for our return to Okaukuejo. Our first full moon was over. Now it rose more than three hours after all light in the western sky had vanished. Sitting out or approaching rhinos in total darkness was suicidal. The night vision scope was adequate with moonlight, but it just couldn't magnify starlight to the point that we could work safely. We'd worry about that problem later.

I looked at Cunningham. Sun and shade flickered across her tanned face. We smiled, savoring our success. No longer disheartened, we radiated enthusiasm on the drive east to Okaukuejo. Sonja chattered. Carol cut mold from cheese with her Swiss Army knife. Even the Long Life milk tasted good. For the first time we felt that we had a study, that we really could gather data, at least in Etosha. Other study areas within the park were still needed, but that too could wait. Below the sweeping sky, woodlands unfolded into grasslands. Hook-thorn acacia, mopane, and nebrownii met savanna. As we drove, my mind drifted to other places, other times, journeys.

I remembered our sweet goodbye to our dogs, Sage and Sika, also our van sputtering and breaking down as I drove to meet Carol and Sonja at my parents' home in Los Angeles before leaving for Africa. Carol's mother and father had come up from San Diego and left before I finally arrived the next morning. All I could do was say goodbye to them over the phone.

"Take care of each other and our only grandchild, take care."

Their voices floated back to me as I drove.

Reality returned as I concentrated on a sharp curve. As we closed the final distance to Okaukuejo, five giraffes crossed the open plains.

The hangar at the airstrip signaled our approach to the fenced enclave, the German tower, the buzz of tourists, the noisy buses, and the local people. Thirty springbok darted from the roadside, their fawn-and-white bodies flashing and moving like a single organism. It was Saturday. The post office would be closed until Monday. We headed for the Institute. Sometimes our mail was there. We were never certain. Anxious to hear from home, we decided to stop.

6

Mom

[Carol]

Although it was the weekend, a few people were working at the Institute. We slipped through the unlocked back entrance into the dim research hall. Low voices rumbled down the corridor. Joel and Sonja stood a few feet away looking at articles and photographs posted on the wall while I sorted through the small pile of mail in the visitor's compartment.

A small yellow envelope caught my attention. "That's strange, someone sent me a telegram."

Tiny cramped handwriting filled just a few lines, but I only made out the first five or six words. JB must have seen the shock in my face. He stared intently at me until I choked out, "Joel, my mom died."

"What?!"

"My mom died."

"Come on," he said, and gently nudged me toward the doors, outside, and away from offices and people.

We drove to the airstrip and stood in the sun crying and hugging each other and Sonja. Soni couldn't figure out what was going on. For a while we simply repeated, "I can't believe it. . . . I can't believe it. Read the telegram again."

We stared, silently, and read the message over and over. How much could a telegram reveal?

She had gotten sick, been hospitalized briefly, and died on the day of the telegram, April 26, 1991. We had been gone just five weeks. Dad's words stared back:

"Sorry for sending the news this way."

7

The Unforgiving Desert

[Joel]

Days passed. Carol's mother had been cremated in San Diego even before we received the news of her death. There was no point in flying back. We headed to the desert, depressed but with hopes of finding rhinos. In Khorixas we stopped to see Blythe and, over a customary cup of tea, we asked about the rhino records that she had shown us during the dehorning operation. As head of Save the Rhino Trust, she coordinated rhino monitoring and maintained records of individual rhinos, horned and dehorned.

"Sharon [Montgomery, in Windhoek] has the records," she told us. "They're being computerized."

So instead of pursuing them, we asked Blythe's advice about possible sites for our desert studies.

Horned rhinos were in the northern Namib, animals dehorned in 1989 were in the central Namib, and the recent dehorning operation had been in the south. We'd been with the operation in the south, but the northern and central regions were new to us. Consulting topographic maps of these remote areas, we drove west following a dirt vehicle spoor. (In this part of the world the word "spoor" has several connotations: It could mean tracks of animals, a dirt road, a trail formed by many animal tracks, even a set of tire tracks crossing virgin desert.) We hoped that this spoor led to a fountain Blythe had described, since we were anxious to find rhino sign on our own.

For two days we navigated boulder fields, eroded hills, and sandy rivers, trying to find the spot she had circled on the map. Carol moved rocks while I drove, then I was the weightlifter. One afternoon our progress was three kilometers. Covered in sweat and caked with dirt, we finally neared a fountain, only to see an enormous structure blocking our path.

Wire between wooden posts three meters high, this was the Veterinary Cordon Fence, which bisects Namibia, stretching from the Skeleton Coast on the Atlantic Ocean all the way to Botswana. Known as the Red Line, it was erected to prevent the spread of disease from wildlife to livestock. Farmers to the north had restricted access to European Community markets; those to the south marketed their goods freely.

The fence also severed migration and killed unsuspecting animals trying to find water during Namibia's frequent droughts. Conservationists Mark and Delia Owens, among others, had battled African authorities over fences and wildlife. This stance had led to the Owens's eviction from Botswana.

Like many of the animals, we too were blocked. Driving parallel to the fence, we found tank-sized holes made by elephants, but we chose not to drive through them. Even if it took several more days to retrace our route, our project was too important to jeopardize by not honoring the barrier. Why Blythe hadn't told us about it was more puzzling. Either she'd assumed we already knew about the fence, or she had been too involved with other things to mention it.

Still, the desert was beautiful. We made a temporary camp above a dry, sandy riverbed lined with mopane trees. A fountain, named !Koibes in Damara, glowed lushly in an expansive plain of red rocks. Bright green reeds grew next to pools. Grasses with feathery seedheads burnt golden by months of summer sparkled in the early morning light. House-sized euphorbia plants, a muted olive green, dotted the landscape.

Unfortunately, we were still unable to find desert rhinos. We tried desperate tactics. Hill-topping, a strategy used by tarantula hawk wasps, was the first. Male wasps simply sit on the highest ridges waiting for females. We did the same. One of us rose before first light, hiked up the nearby hill, and, while munching stale biscuits, scanned the desolate landscape with the spotting scope. If we saw a rhino, the plan was to wake the other person (and Sonja), hike, and, without disturbing the animal, get a photo at close range.

From the hilltop we scanned 10 to 15 square kilometers of desert plains, flat expanses broken by deep gullies and eroded mountains. Hundreds of springbok, bands of mountain zebras, and half a dozen giraffes fed on distant horizons. Gemsbok came to drink. Ostriches shimmered on faraway slopes. We didn't see any rhinos.

Sonja was learning more words. "Bok" meant springbok and "Ino" was short for rhino, the latter known mostly from pictures in our southern African mammal guide. This book sat in the front of the Land Rover, as did a bird guide, a compass, toothbrushes, and a Swiss Army knife. In the storage box between the front seats and scattered on the dash were maps, water bottles, sunscreen, Labiosan for nose and lips, children's books, and cassettes. The Land Rover was home and office. We needed everything.

"Joel, wake up! Do you hear that?" Carol whispered.

I awoke with a start inside the pitch-dark tent.

"Sshhh, I heard something."

Despite being rare in Kunene Province, a small pride of 10 or so lions lived to the north. Since lions generally need to see movement to stimulate their predatory actions, we expected to be safe inside the tent. Now, neither of us shifted a muscle. We lay listening. I pressed the button on my Casio watch: 4:20 A.M. I tried to be calm. I drummed up facts about lions. *Most active at night; hunting efficiency increases with darkness. Females better hunters than males.*

"During dark nights you could hear them [lions] panting outside the kraal, but as soon as the moon appeared they were gone," wrote Peter Moller in his 1899 book, *Resa i Afrika genom Angola, Ovampo och Damaraland.*

Friends had similar observations: "We watched lions stalk a wildebeest on a partly cloudy night. When the moon appeared, the lions lay down and disappeared in short grass. As soon as clouds covered the moon again, the lions were up and hunting. It was near the very water hole in Etosha where you saw your last rhino."

Now there was no moon. It had set three hours earlier. Soon it would be my turn for early morning hill-topping. The prospect of hiking up and down mountains excited me less and less. "Whooop, whooooop" echoed down to us, a spotted hyena prowling somewhere far from !Koibes. We both lay silent, wondering what was beyond the thin tent canvas.

We woke up groggy from the heat, the sun beating against the tent. It was 7:30 A.M. Soni was stretched out on top of her blanket next to Carol. I poked my head out the front door.

"Kwaak, kwaak." Pied crows had gathered at last night's dirty dishes. Now they called in alarm and flapped their wings, rising white and black against a blue sky. Outside, footprints stamped in the red soil led directly to the tent and then turned abruptly. The distance between each print increased, and they grew deeper. The animal had been running. Our predawn visitor had been a solitary elephant—a bull no doubt, judging by the prints' huge size—surprised by the tent in its path.

As days slipped by, we discussed other options for finding rhinos. If we didn't do things differently, we'd end up with no data.

"Maybe we should try observations at night. They worked well in Etosha. The moon is large enough, and with the night vision equipment, we should be able to see if hyenas or elephants come. What do you think?" I asked.

"Sounds okay to me. Do you think we can find a spot where we can set up on the roof rack and still see the water? Those rock walls above the spring are pretty steep." Carol sipped hot Roibos tea in the sweltering afternoon heat.

"Forget the car. Let's build a hyena-proof fort and sit in it."

"Great idea. You sit in it, I'm not."

Five hours later, the rocks had been hoisted and set neatly into place. The rounded structure, about two meters across and one meter high, offered visibility in all directions. It rested on a large rock slab above a miniature canyon and seemed an unlikely place to encounter elephants or rhinos, although a steep ledge near us would provide an escape route, if we needed one. From the fort we could also see the Land Rover, should Sonja need us. A silent cotton blanket replaced our noisy high-tech jackets.

To persuade Carol to join me in the fort, I made dinner. She appreciated the gesture, but her smile was brief. The gelatinous red wads of canned Bully Beef over rehydrated peas and rice looked terrible. They tasted worse. Sonja spooned down half a bowl, but Carol and I only managed a few bites. Stale crackers were preferable.

About an hour after dark, a band of mountain zebras appeared. They bunched and shuffled nervously near the water. Suddenly two hyenas materialized from nowhere, then a third. In erratic bursts, they approached us. Twenty meters from our fort, one plopped down and stared. We didn't blink. Another came to within five meters, then three. When it paused, sniffing vigorously, I stood up, startling it and scattering all three. Even the zebras bolted from the hillside. The night's excitement had ended.

By midnight, cool winds were blowing in from the Skeleton Coast. Carol pushed against me for extra warmth. By 2 A.M., eyes glazed and head nodding, I realized I couldn't fight sleep any longer. Both of us too tired to keep watch, we gave up, walking the 200 meters to the Land Rover and climbing onto the padded roof rack where our breezy open-air bedroom waited. Early the next morning, I searched the muddy edges of !Koibes to see if rhinos had drunk while we slept. It had been another discouraging night. No rhinos had come.

A quiet hum caught Carol's ear, different from the familiar drone of our small refrigerator tucked into the Land Rover. The sound grew stronger. We looked up, expecting a plane. Bursting over the ridge, it was upon us before we knew it—a Land Rover filled with young native boys. I stepped forward, offering a friendly wave. This was the first vehicle we'd seen in the desert.

A tall man, bearded and in tattered pants, climbed out. He looked familiar, but we couldn't place him. He hadn't been at the dehorning operation.

"Hello. I'm Garth Owen-Smith."

Now we knew why we recognized him. His conservation achievements had been lauded recently in a *Time* magazine feature entitled "The World According to Garth."

Owen-Smith got local communities involved in conservation. People living on farms or villages in remote areas were given responsibilities for monitoring local wildlife. The concept of "game guards" worked well, although the term guards, in the westernized sense, was too strong. Community game guards were local trackers, designees of tribal chiefs who walked patrols in exchange for food or small salaries. Most were unarmed and not uniformed.

Years earlier, Garth had been commissioned by the New York Zoological Society to study methods of censusing desert rhinos. This man knew the Kaokoveld's 80,000 square kilometers, its people, its politics. We had first heard of him through his brother, Norman, who had studied white rhinos in South Africa for his doctorate. We had tried to contact Garth from my base at the University of Nevada in Reno about study logistics, but he had never received our letter. He was unaware of Americans about to study Namibian rhinos.

Garth invited us to the environmental and community center at Wereldsend (World's End), a facility that he and anthropologist Margaret Jacobsohn had developed to further conservation in the northern Namib Desert. It lay on a dirt spoor about four hours south of us, 150 kilometers west of Khorixas.

Over the next few days, we tried other ways to find rhinos. I tracked footprints through thickets, on flood plains, and in dry riverbeds. The evidence for rhinos was uncontestable, fresh droppings. But after encountering an elephant in a steep canyon, I knew that something Garth had said was right: "Tracking alone is a sure way to be killed."

During my explorations, Carol and Sonja stayed with the Land Rover, waiting patiently above the waterless rivers. We talked on hand-held two-way radios. Batteries were always a problem, so we restricted our conversations to three minutes at the top of every hour. The NiCad radios never charged fully from solar panels we had brought from the states, so we ran electricity from the Land Rover's 12-volt system with an invertor that changed the current to 110 AC.

The days seemed to fly past. We moved camps, looking for rhino sign near other water holes. Each shift took another day to pack and part of another to unload.

Other than the Land Rover, we would have no home. A small caravan at Okaukuejo offered a nice break but we would use it rarely, about 10 days every two months. Camping got old. We were always hot and dirty. The water we carried was primarily for drinking, much too valuable to waste on bathing. Once our supply was gone, it would take days to drive somewhere to replenish it. Our appetites left us. Although the scenery was beautiful, and rhinos lived there, failing to find them was discouraging.

Wereldsend was situated in a lovely valley, about an hour east of Skeleton Coast National Park. Depending on the rainfall, the adjoining plains might be thick with knee-high grass or just dust and gravel. The presence of springbok, gemsbok, and ostriches told us that this was a good year, and food remained.

Margaret Jacobsohn, trim and dark-haired, greeted us. She had just written an acclaimed book on the Himba, nomadic semipastoralists of the northern Kaokoveld who were often compared to East Africa's Masai. Our initial conversations with Margie and Garth were less than auspicious, although good-natured. They chided us about America's global influence and unregulated consumption of the earth's resources. We couldn't disagree. So we talked of their South African ancestry and apartheid. The four of us sparred over culture, ideology, and fairness. We also focused on Garth's work on rhinos and that of his brother Norman.

A tour of the center revealed rows of chalk-colored rhino skulls. Many were scarred, their nasal bones shattered, with holes where pangas had hacked below the horns. Garth knew most of the rhinos by name; he had known them alive, before the poaching epidemic began. The morgue, shaded by the only trees for miles, overflowed with elephant tusks, kudu horns, and the skulls of lions, giraffes, and baboons. These were chilling reminders of humans' indifference to the natural world, and the center used them to teach others about the diversity of life in this wonderful ecosystem. Two Brits, Tim and Rosie Holmes, were the education officers, committed, bright, and seasoned by experiences in Uganda.

As day faded, we listened to the music of Toni Childs and Tracy Chapman, feasted on pasta and wine, and discussed biology and conservation, anthropology and politics—the ingredients of life. We climbed to bed happy, on our rocking Land Rover that we jokingly christened the Boat. We contemplated our research goals and our inability to find rhinos in the unforgiving desert, still discouraged but thinking there might be hope. If others had been able to find rhinos in this area, we could too.

Garth and Margie interrupted our planned departure the next morning, suggesting that we wait an extra day. They'd go with us to find one of the rhinos that had been dehorned two years earlier. Garth would show us how to track.

We all piled into their Land Rover, the three of us, the two of them, and two stout sweet Staffordshire terriers, Chaba and Kumbo.

A lappet-faced vulture perched in a lone tree, capturing the warmth of the sun. With two forceful strokes, it powered itself off its branch and floated on the thermals. Garth pointed to tracks of hyenas and a set he thought were cheetah in the Springbok River below. Welwitschia, one of Namibia's endemic plants, reached deeply for traces of moisture, two large leaves curling out from a woody stem. Distinctive and often ancient, individual plants may live up to a thousand years.

We found more elephant spoor in the dry river bottom. "These animals have good memories and can be dangerous. It wasn't that long ago that they were poached. Be careful," advised Garth.

We checked dung at frequent intervals—piles of it. By stepping on the darkened patties, we determined their freshness. Wet, recently deposited dung meant the animal was surely close. Garth showed us how to tell zebra hoofprints from those of poacher's donkeys. The hooves of mountain zebras are worn down by the coarse rocks until the V-shaped pad in the middle of the sole made no impression on the ground. In contrast, the pads of the donkeys were clearly visible. Shoeless humans were obvious by their tracks. Garth explained that both signal the possibility of poachers.

Around noon we found rhino tracks that looked as if they had been made minutes earlier. Garth led the way on foot. Margie, Carol, Sonja, and the two dogs stayed in the Land Rover, following slowly half a kilometer behind us. On open plains with no place to escape, Garth's formula for safety was a nearby vehicle. Blythe had mostly abandoned hiking, and many of her observations were made wisely from a vehicle.

When the vegetation thickened, Garth skirted it widely, always working with the wind's direction so that the rhino wouldn't smell us. Around each corner I expected a rhino, but none appeared. Carol and I traded places so that she could track with Garth. At one point, he picked up a masticated welwitschia leaf. Despite the midday heat, it was still moist. The rhino must be close. I joined Garth again for the finale.

Some 400 meters ahead, a single Boscia tree grew off the gravel slope. A gray mound lay at its base. In the sole shady spot in a broad valley was a sleeping rhino. A rhino. We had tracked a rhino. Actually, Garth tracked the rhino. We only followed, but that detail didn't bother us. This was a data point. A datum. Our first in the desert.

It was a dehorned desert rhino, a female that Blythe Loutit had named *Mystique*.

We had learned more about rhinos and finding them that day than we had during the two weeks spent with the dehorning team.

Rejuvenated by Garth and Margie's hospitality and injection of encouragement, we left Wereldsend and headed back north. Still nagging was Garth's comment: "If you want to succeed, you'll have to find a tracker." We needed someone who would work with us and live with us, someone who knew the area.

For the time being we very much wanted to see if we could find rhinos in an area of higher density. Ecologically, density is a relative thing. If a habitat has greater plant productivity, then it will support more animals, but it does not necessarily follow that each individual will have more food per se.

The area that we now traveled to, Aub Canyon, had similar habitats but was more wooded than the areas we had so far visited. The desert there was still very dry, receiving less than 75 millimeters of rain yearly. It reminded us of Death Valley—large, expansive and scorched. But somehow there were giraffes, elephants, lions, and zebras. It was also where Garth had seen 23 rhinos in just nine days.

Eugene Joubert's comment that we'd be doing fine if we found one rhino per week still haunted us. Garth was unusually familiar with the area and splendid in the field. From our experiences so far, we now understood why solid progress meant four rhinos a month. Nevertheless, we desperately wished to exceed Joubert's expectations. At this point we were losing. In three weeks the only rhino we had found was a present from Garth.

8

A Tracker Appears

[Joel]

We inched along, on spoor above Aub Canyon, passing through hills burnt brown by heat, our cache of food and water dwindling rapidly. Mighty streaks of sunlight pierced the clouds. This was day 23 for us. Green trees lined the gullies as the two-track spoor veered between steep slopes. Music played. Raffi was singing for Sonja:

"Down on grandpa's farm, there is a big brown cow. . . . the cow, she makes a sound like this, Moooo, the cow, she makes a sound like this, Moooo. Down on grandpa's farm, there is a. . . ."

"A rhino, *Rhino*, a RHINO," shrieked Carol.

"Where, *Where?*" I slammed on the brakes.

"Ino, Ino." Sonja saw it at the same time.

There it was, no more than 200 meters from us, a mere 50 meters off the road. What an opportunity. We had the car for safety, and the rhino was so close.

Its ears were erect, its body stolid; it had detected us. It would now be difficult to sneak within the 70 meters needed for a data photo.

The Leitz rangefinder, buried somewhere in the Land Rover's cargo area, was critical for distance measurements. If the gap between the rhino and me was known, its horn's size could be estimated with 95 percent accuracy just by scaling the photo on a computer. But without locating that rangefinder, I'd be forced to use the other. Like the Leitz, the Mitutoyo was accurate, but the distance between the subject and photographer had to be much closer, less than 35 meters away, a proximity that beckons an attack in the daytime. The Mitutoyo was for night work. The Leitz, on the other hand, in conjunction with the 500-millimeter lens was for day work. While it was still risky, the extra 35 meters might offer a narrow margin of safety.

The rhino continued standing, waiting, as if inviting a showdown. Frantically, I searched the cargo compartment. During the drive, all the bouncing had caused the refrigerator to break lose, knocking two trunks aside and smashing the food boxes. Somewhere in the jumble lay the Leitz rangefinder.

I took too long. The rhino bolted and vanished. The fleeting view offered anatomical information—a cord leading to the scrotum. The animal was male. It was time to forget the damn Leitz. I grabbed the Mitutoyo rangefinder, knowing that I'd now have to be dangerously close for an identification photo and accurate horn measurements. Before Carol knew it, I had disappeared over the edge of the ravine.

I didn't track the rhino. I was too afraid to remove my eyes from the thicket where he had disappeared. My steps were precise. I avoided gravel. Large sturdy rocks were quiet, pebbles made sound. I eyed trees to locate climbable branches, noting the width of each tree-trunk. Sweat bubbled on my forehead and trickled down my temples. The world was silent. I waited minutes, then took several more steps, and waited again.

The rhino had not gone far. Anticipating my approach, he pressed his rump deeply into a giant euphorbia, his ears forward. His head had strikingly large horns. His left ear had a deep tear and was more tattered than the right.

I focused the camera. He looked mean. I had never seen a rhino up close in the daytime. The Mitutoyo indicated exactly how far the telephoto lens had been extended, a mere 1.21 millimeters. My heart sank; he was still 80 meters away. I'd have to move another 50 meters closer. I couldn't tell if he saw me. A mopane tree lay 30 meters ahead. Its branches were tall and offered safety. If only I could reach it. Every step crunched. The rhino's ears swiveled. At this point only 44 meters separated us. I passed the last tree and continued my assault. Only a few more meters separated me from the point where I could take the priceless shot. Data. *The photo!*

The rhino stepped forward then stopped. I was close. Dry leaves blocked my progress. I was immobile, not wanting to retreat but afraid to go on. A sophisticated sensory apparatus, the rhino's ears, honed by millions of years of natural selection, was locked on my every move. I waited. Three minutes passed. Mopane flies crawled on my face and crept into my eyes for moisture. I had to shift positions.

A noise, something behind me, broke my concentration. I dared not move my eyes from the rhino. I heard something lumbering toward me, careening in the riverbed . . . the Land Rover, swooping in my direction, bouncing over rocks. It was Cunningham! In a fraction of a second, the rhino pivoted and disappeared, climbing up and over the ravine's edge. I ran after him, trying to get to the top to see. It was fruitless. He was gone. In a few precious seconds he had evaporated.

"What were you thinking? You blew it. Was that on purpose?" I fumed, trying to understand why Carol would deliberately scare the rhino when only a few meters remained until I snapped the photo.

"I thought you were in trouble, that you were stuck and couldn't get to the tree."

She looked shocked that I was angry.

"You must be completely paranoid. I wasn't in trouble, just waiting for the right moment. These things take time!"

"Fine. Find someone else. If you don't want to work with me, I'll go home." Tears streamed down her cheeks. Minutes passed. We didn't speak. After a while, I hugged her, and we made peace.

Later she explained. From the top of the ravine she could see the rhino, not me. When I had finally appeared, I looked stranded. The tree I thought was close to me she saw as too far away. The rhino and I looked due for a nasty collision. She also had thought I had already shot the photo, because I had spent so much time looking through the telephoto lens.

"It all depends on your perspective," she said, as if hoping I'd see her view.

In the fading twilight we made a hasty camp, put Sonja to bed, and sat under the stars. Warm red wine from Stellenbosch, South Africa's wine country deep in the Cape, helped us relax. Lions roared somewhere to our south. We watched shooting stars. The strain of fieldwork was affecting us.

More days passed. We explored thousands of square kilometers. We saw fresh elephant dung, a dead kudu, bones cleaned by jackals, and a long black horn sheath once carried by a proud gemsbok. Chips of ivory were abundant, left from elephants that broomed their tusks while digging or, perhaps, fighting. At night the unmistakable roars of lions close by kept us awake.

One morning while walking on the rock shelf above a fountain to see if animals had left spoor from the night before, movement on the ground caught our attention. A long slender snake, olive-brown and quick, raced off. A black mamba. These formidable snakes eat small birds and mammals, pursuing and then stabbing them with their fangs until neurotoxic venom paralyzes their muscles. Death occurs from respiratory failure. We began showing Sonja pictures of "slangs" (snakes) and told her to move away and call us if she ever saw one.

Our energy was draining. We wanted fresh greens, a tossed salad. Our last had been in the States. A diet change was needed. We wanted bread, even meat. We talked of enchiladas, burritos, tacos. Our appetites returned. We made ourselves sick with thoughts of savory dishes, talk of movies and libraries. We discussed omnivory and carnivory. Humans needed protein. Baboons and chimpanzees ate it now and then. We wanted springbok steaks.

Fresh meat was not a real option, so we decided to make bread. Neither of us had ever done so. Like most Americans, we got everything we needed at the corner grocery. We chastised each other for our lack of culinary skill.

As Cunningham evicted bugs from our flour and pulled dry yeast from a supply box, Soni and I gathered wood for a fire. We thought we needed oil to moisten the batter, so we drained a can of Pickled Cape Fish into mealie (corn meal), and added beer, sugar, and dried milk. Sonja stirred the goop with both arms. We sandwiched the cast-iron pot in a bed of hot coals and moved to the tent for shade.

In minutes a burning smell wafted in our direction. Quickly knocking glowing embers from the lid, we decided that next time we'd cook longer and with fewer coals. After we sawed off the hard black crust, a warm and faintly fishy flavor greeted us. It was delicious.

That night, near a fountain, we tried another all-night watch, our ninth so far in the desert. No luck. Only twice so far had we succeeded in finding desert rhinos at night.

Our stomachs now controlled us. We tried a different mixture for the next day's bread. This time we extracted oil from Black Cat peanut butter and mixed it with dried raisins. That loaf was nearly as burned, but it too was tasty. Again we sat up at night, waiting for elusive rhinos. Predictably, none came. The next day, more bread. A routine was developing. This time, we attempted to get oil by frying biltong like bacon. No success. The kudu must have been dreadfully lean.

The routine was short-lived. On day 30, less than 30 liters of water remained. If we experienced car problems or were stranded, our problems would turn grave. Juices and canned fruit had been gone for nearly a week. Rehydration might not be an option. It was time for a trip to Khorixas for supplies.

Leaving the area, we noticed white and black feathers covering a stream bed east of the mighty Uniab. Dry and wide, this river offers refuge to wildlife seeking relief from the inferno of the gravel plains. The disheveled mass of green kori bushes suggested a desperate struggle. Earlier, at 4 A.M., a chorus—"whooooop, whoo-oooop, whooo-oooop"—had continued on and off until near dawn. This sinewy ostrich would run no more.

Spotted hyenas are first-rate carnivores, large and dangerous. In the northern Namib Desert, they form small groups, unlike the more familiar large throngs of East Africa or even the western Kalahari. In this part of the Namib Desert, spotted hyenas still kill gemsbok and mountain zebra. Each clan member may eat up to eight kilograms of flesh nightly. We had seen five hyenas together and, less than an hour later, what we presumed were another three, 25 kilometers away. Although spotted hyenas had been studied by Joh Henschel and Ron Tilson in the central Namib and by Bill and Kathy Gasaway in Etosha, little information existed about them in our desert areas.

More important to us, there was little knowledge about their relationships with either of Africa's two species of rhinoceroses. Maimed calves of black rhinos had been described in the scientific literature for nearly 30 years, first in Kenya and Tanzania and later in parks in South Africa. Originally, calves missing ears were thought to be mutants, genetically impoverished. Later, however, it became clear that calves who had been known as individuals with intact ears or tails had subsequently lost them. Although lions may be responsible for the maiming in some regions, calf maiming has not been reported from areas where spotted hyenas do not occur. We had a healthy respect for these primal carnivores, particularly after hearing their powerful jaws crushing elephant bones months before.

One of the prime biological arguments for dehorning Namib Desert rhinos was that predator densities are low. While we had yet to gather rhino data, a 1991 study in Kenya's Aberdares Mountains had revealed that more than 40 percent of the calves had scars or were missing ears or tails. The density of spotted hyenas there was higher than most areas outside of Ngorongoro Crater in Tanzania. Calves had also been maimed by hyenas in South Africa's Umfolozi-Uluhluwe Reserve, but neither maiming nor hyena density was as great there as in the Aberdares. If hyenas affected the survival of rhino calves in areas where mothers were horned, we reasoned that they certainly might affect calf survivorship in areas where mothers were dehorned. The important variables here concerned the availability of alternative prey for hyenas. For us, immediate questions were more critical. Could we find rhinos? Could we determine if they had calves? If so, we could say something about calf survival to a specified age.

In Khorixas we again sought Blythe. Although she had not yet retrieved her binders with information on rhinos ear notches from Windhoek, there was good news. She suggested that we hire her former tracker, Archie Gawuseb.

Archie, whose Damara name was !Huie, had been convicted of rhino poaching. Most days he sat idly in the Save the Rhino Trust headquarters, a small, crumbly building on one of Khorixas's dusty streets. He had been one of Blythe's star trackers until he was led astray by alcohol and bad company. Blythe told us that he had four years' probation, was still under the guardianship of Save the Rhino Trust, and was required to perform service for a conservation project. He could no longer track rhinos for Save the Rhino Trust, because he was no longer trusted by the other trackers. Blythe said he could be assigned to us.

Swell. A convicted felon. Bad company. Several Afrikaners warned us about hiring "one of those chaps."

There were logistical concerns. Who would feed, outfit, and transport him to and from his Sesfontein home, nearly a full day's drive away? Our car was already overloaded. Where we would put an extra person, and all his food, water, and other supplies?

Blythe's husband, Rudi Loutit, the head conservator for Damaraland, confided that Archie was to give key testimony in a forthcoming trial against another poacher.

"He must be kept with you for safety, because his life might not be worth much in Khorixas, or even in Sesfontein."

There were death threats against Archie because of his plan to testify. A spell had already been cast on him by a local witch doctor, and a spell would also be cast upon all those who helped him. That meant us. We didn't believe in witchcraft, but locals took it seriously. We didn't know what to do. We weren't equipped for serious protection. Archie's best defense was the remoteness of our desert locations. At least he knew a little English.

Blythe suggested that we try him for a month; she'd supply his food. All we needed to do was pay him what Save the Rhino Trust offered their trackers, 250 rand ($100) per month. We agonized over the decision. Would Rudi try to have our research clearance canceled if we didn't agree? With reluctance, but little hope of finding rhinos on our own, we agreed. We'd all meet at Wereldsend in a week.

We waited there on the seventh day, and on the eighth. Archie never arrived. We left, dejected and hoping that we'd be able to find desert rhinos on our own. At least Bill and Kathy Gasaway planned to take a short break from their project on Etosha's ungulates and carnivores and meet us. But despite our best efforts at tracking rhinos, the results were the same. After a few kilometers, the spoor always disappeared.

One early morning Cunningham spotted a rhino in the hills near our camp, and we got to within 150 meters. Like the earlier male, this rhino backed into a bush. For ten minutes he stood frozen as if deciding whether to attack or run. Then he trotted into a side canyon, heading up and over rugged mountains.

Reports of rhino behavior differ widely. "Naturally timid and certainly not dangerously aggressive" was how V. F. Kirby described them in Frances Selous's 1908 *African Nature Notes and Reminiscences.* Carl Akeley, the title of whose 1925 book *In Brightest Africa* played on Henry Stanley's 1890 *In Darkest Africa,* wrote of rhinos: "It is also true that as soon as he smells man he is likely to start charging around in a most terrifying manner." Earlier, in 1856, Charles Andersson wrote in *Lake Ngami:* "They are . . . of unprovoked fury, rushing and charging with inconceivable fierceness animals, stones, bushes—in short, any object that comes their way."

Rhinos were curious. Erratic too. We didn't know whether they'd run or charge. It was no consolation that others couldn't help to explain this unpredictable behavior. Why rhinos are afraid is unclear. As adults they are not preyed on by lions or hyenas. That females with calves are vigilant makes biological sense, but why males run was something we couldn't explain. Perhaps the explanation was as simple as the fact that rhinos were now being slaughtered by humans. They flee when at all possible. But that rationalization was troubling. The historical reports of vigilance and flight predated the recent wave of wide-scale poaching. Maybe rhinos were just programmed to be skittish, an idea that makes little evolutionary sense.

Our search for rhinos continued. Carol remained with the Land Rover and Sonja, reading and enjoying a thin sliver of remaining shade. The Gasaways and I followed spoor.

At the top of the hour, Cunningham and I talked on the radio. A truck with four black men had approached her—only the second time we had encountered humans in the remote Namib. Archie was abroad. He had no equipment. No tent, no binoculars, not even a hat. His arsenal consisted of tattered boots, one small container of water, a paper bag with rancid butter, and a mattress.

We were bewildered, and Archie offered no explanation. Two days passed. He was quiet, except that, before retiring for the evening, he'd say good night. Even with Archie, we had still not found rhinos. Archie remained silent. We wondered whether he was naturally reserved or if he was reluctant to respond to Bill's eagerness to learn more about him and the area. We were all intent on knowing more about this man and his poaching, but we deliberately avoided asking, trying to respect his privacy.

On the third day, I asked Archie if he had other clothes. He didn't. I gave him underwear, socks, and my blue Nike hat. Archie explained that he'd been told that he'd track for us for only a day. At least that was his understanding. He had no idea that he was to spend a month with us, or that bags of mealie would be provided, or that he would be paid by us. He knew absolutely nothing of the arrangements that had been made about him. Incredible.

We asked if he was interested in working for us, being part of our team, to help us study rhinos. We explained our project, who we were, why we were in Africa, and what we had done in the past. (During the dehorning operation, none of the trackers had been told that we were in Namibia to study their rhinos and dehorning. No explanation had been offered about the guy who rode with the trackers for days in the back of the bakkie, or why he measured the bodies of immobilized rhinos and opened their mouths and examined teeth. I had tried to explain my purpose, but language barriers had interfered.) Because Archie spoke some English, he understood when we told him that if he decided to work with us we'd pay bonuses for each rhino that we photographed at close range. We'd teach him more about cameras and English. In return, he could teach us about tracking and rhinos. Our optimism soared.

Days later it crashed. We had not found a single rhino, although we crawled along, driving up to 70 kilometers a day. We tracked through more heat, across gravel, up and down mountains, and in thick riverbeds. We began to wonder if Archie was overrated. Perhaps he couldn't really find rhinos. We had been warned that "those chaps were lazy" and a good deal of variation existed in their abilities.

We desperately wanted to believe an alternative explanation, that our poor success resulted from too many bodies trailing behind Archie, unable to match his pace. We were completely inexperienced at tracking and in its subtleties; perhaps we impeded him. It was common for Archie to track alone, maybe with one other person, not three. Archie traditionally just walked. No water, no food, no pack. He sailed over rugged landscapes. If a long day were ahead, he'd fill a plastic Coca-Cola bottle with water. He was similar to the other trackers we met. In contrast, we Americans were encumbered. We needed sunscreen, water bottles, and lunch. We brought along two-way radios, a first aid kit, and a data book. Then we added the heavy camera, 500-millimeter lens, a monopod, and a rangefinder. Archie was patient. He never said the obvious, that we delayed him.

Bill, who had worked in Africa on and off for the past five years, was a respected international scientist with a keen interest in conservation. He understood the vagaries of fieldwork and appreciated the need for large samples for statistical analyses. Well-acquainted with studies of demography and ecology, he suggested that work in the desert was likely to produce little data. Perhaps our time could be better spent focusing on Etosha's rhinos. After all, how many days could we spend searching and searching without success? Could we really decipher much about desert rhinos given how rarely we found them?

Maybe Bill was right. This might be futile, even with Archie. The tracking and dehorning ventures near Doros Crater had taught us that even with 8–15 trackers, a helicopter, and four or more vehicles, finding rhinos was difficult. Once Bill and Kathy left, there would only be three adults and a baby. We pondered the possibility of defeat and considered whether we should remove the desert from our study design. But if we did, there would be no data on dehorned rhinos.

Frustrated, we prepared to return Archie to Sesfontein. We broke camp, hoisting jerry cans now empty of petrol, trunks, and water jugs onto the Land Rover's roof. Cabling everything down took 30 minutes. We now carried three spare tires. It was going to be yet another day of intense sun, shimmering heat, and grinding over gravel.

By midday we had found a two-track spoor leading south. We were unsure where we were, but we weren't concerned. Gemsbok and ostriches broke through the haze. The "quark, quark, quark" of a Ruppell's khoraan, with its delicate head and neck markings, and the sight of giraffes enriched the monotonous driving.

We passed through a small gorge where walls of red rock obscured our view. Sonja sat between Archie and me, balancing on the storage box. Carol had crammed herself into the back seat, her legs trapped by four empty 20-liter water jugs. Archie selected a Sergio Leone tape. Tunes from Spanish desert scenes in the Clint Eastwood film *The Good, the Bad, and the Ugly* trumpeted forth.

"Ino, ino", Sonja called out as we emerged from the gorge, and she pointed toward boulders on a faroff hillside. Carol craned her neck, peering from behind us.

"I don't know, Joel. I don't see anything. Maybe she thought those big rocks were rhinos."

Then came a frenzied honking as the Gasaway's small pick-up sped up to us from behind. This was not their typically cautious driving. Careening to a stop, they leaped out, yelling, "Rhino, rhino. Didn't you guys see them? Two rhinos came out of the gully and crossed right behind you."

"Where, where?" I scanned the hillsides frantically. We hadn't driven more than 200 meters since Soni had spotted them. I couldn't believe that we hadn't believed her. She was too young to make up stories.

I threw the gearshift into reverse and powered quickly back to the gorge. We speculated about Sonja's sighting. The rhinos had not passed in front of us, so she must have seen them in the right rearview mirror, which explained why she had pointed at the hillside.

Despite the rocky substrate, Archie found the fresh spoor instantly, a mother and a baby. We couldn't see the animals, but we knew they had been here, running, kicking rock, and denting the earth with hard toenails. Archie and I each grabbed a camera and lens, equipment now kept within arm's reach. Quickly we traded sandals for boots and were off. Pink Labiosan protected my nose from the sun. Archie didn't need any and always smiled when I offered it to him.

Rolling hills, distant buttes, and giant pliable euphorbia spread for miles. If the rhinos decided to charge, we'd be in trouble; there were no trees. Forty minutes later we returned, jubilant. I raised my open hand and slammed it onto Archie's palm. He had no idea what was going on. Exchanging "fives" was not part of Damara or Namibian society. But he reciprocated. His hand smacked onto mine. We smiled and kept smiling. The day was wonderful. We had data.

"Great job, excellent tracking," I said.

We had photographed *Tina* and her calf, which we named *Tiny*—perhaps four months old. Her puny spoor was 13.5 centimeters across. Her mother's was 21 centimeters. Our nomenclature system almost always used the first few letters of a female's name for that of her calf, making it easy to know which rhinos were related.

Garth Owen-Smith had kept track of local animals, and *Tina* was a cow he'd known for a long time. Her anterior horn had been broken nearly 10 years earlier, and had now regrown completely. We decided to make duplicates of our close-up photos to update Garth's files.

Earlier Blythe had sent us a letter, with copies to Malan Lindeque at Etosha and to Rudi, her husband. It concerned data and photos:

> Joel, if possible we would like to have a properly drawn up programme of your intended pattern of work. . . . Rudi is the Chief Conservator responsible for all projects . . . it is necessary for us to know your movements as you are dealing with an extremely rare and valuable animal. I would also like to have exact plotting of your rhino sightings, plus sex and estimated age and a copy of each photographed animal as a continuation to our rhino data. . . . Best Wishes, Blythe.

It certainly made sense that Blythe too should have copies of our photos. Once Carol was finished in the darkroom, we mailed her photos of the three animals to date—*Tina, Tiny,* and the bull from Aub Canyon.

Janet Rachlow, one of my Ph.D. students, would soon arrive at the airport 70 kilometers east of Windhoek. Like us, she had never been to Africa. She was also coming to study the biological consequences of dehorning, but her work would be in Zimbabwe and on white rhinos. Janet was stopping over for a few weeks before pushing on to that country's finest national park, Hwange. Zimbabwe, like

Namibia and Swaziland, was involved in last-ditch efforts to prevent their rhinos from being poached. Horns were being removed from both species, black and white.

The night before Janet's arrival, we camped in thick thornveld near the airport. Sonja slept soundly in her usual back-seat spot. Her tiny body nestled above water jugs, duffle bags, and food, all tucked into the leg space below her to prevent her from falling. On the roof rack Carol and I cuddled together for warmth under a chilly sky filled with stars.

At 2 A.M. Carol woke up suddenly.

"Did you hear that?"

"What?"

"I heard something. Was that Sonja?"

We listened but heard nothing. It was cold, below freezing. This was Namibia's winter. The central highlands on both sides of Windhoek regularly dip below 32°F. Occasionally it snows, more often in the hills to the north of the Orange River, the common boundary with South Africa.

Carol peeled back the sleeping bag and climbed in her panties down the ladder. She peered through the closed windows. Then she opened the front door and, reaching over the front seat, patted the back seat blindly. Blankets were there, but no Soni. She was gone. Carol ran to the other side of the car and opened the door. A tiny body, legs in the air, head out of sight, started to topple out. Sonja had fallen off the seat, her head wedged in the paraphernalia.

I awoke to a flurry of activity and hyperventilation.

"I can't believe it," yelled Carol, half hysterical. "What if I'd assumed that she was okay and had just banged up against the door? What if she'd suffocated?"

We both wondered how she would have fared uncovered and in the frigid air. It was still hours until morning.

Janet arrived with radio-telemetry gear and financial support from the Frankfurt Zoological Society. We introduced her to Eugene Joubert. Over beer we talked about conservation, South Africa, and the world's media.

"Yes," said Eugene wryly, "all South Africans are racists," responding to portrayals of Afrikaners. He was kidding of course, but his major point did not go unnoticed. Generalizations about any society can be dangerous.

Janet's first task was to find a sturdy vehicle and get it to Zimbabwe. Reliable, newer vehicles were easier to find in Namibia. We helped. Our bank account had finally been established, so we traded Janet's American dollars for rand. Twenty thousand rand was a good bargain for a 1977 Land Rover. Disturbingly, the largest currency in the bank was a 50-rand note. More than 400 individual bills would have to be counted. We asked the clerk for a security booth to count the money.

"No."

All of it had to be counted right there, in front of a long line of watchful eyes— every single bill. Janet would then have to carry the money through crowded streets, bumping bodies, and fast hands, to our Land Rover, which was parked a kilometer away. So, we all went together, Carol on one side and myself on the other, with Janet and Soni in the middle.

Ten days later Janet and I drove toward Zimbabwe, heading north to the Angola border and across the fertile Caprivi Strip. Tall forests of kiaat trees, Hyphaene palms, and fields planted with mahango filled the flat landscape. Large rivers with crocodiles and hippos flowed north of the road. Refugee-filled towns and police checkpoints slowed our progress. Officials asked us about weapons and bombs, then elephant tusks. Were we travelers, terrorists, or smugglers?

Several border hassles later, we entered Zimbabwe. Along the way we had seen lions, elephants, ground-dwelling hornbills with red faces and throat patches, glossy ibises, and yellow-billed storks. Janet's project had been launched. I caught a flight from Harare back to Windhoek.

Carol and Sonja were waiting. That night we slept in sweats and socks, awakening to a thick layer of frost covering our sleeping bag. Soni was bundled snugly inside the very carefully packed back seat with her dolls, Ted and Marianne. The early morning temperature was a chilly 17°F. The veld was still dry. Rhinos would need water. It was time for more data.

9

It Depends on Your Perspective

[Carol]

We returned to Doros Crater and the Ugab River. During the next few weeks we would try to find rhinos on our own. Here rhino densities were lowest, the heat most extreme, the tracking most treacherous.

Volcanic extrusions, blackened domes, and canyons brindled with ancient gneiss offered dreary reminders of past days with no rhinos. Our first camp was in the bed of a canyon we named San Cave, the spot with Bushman paintings that we'd hiked to when we first arrived in Namibia. Joel and I unloaded the Land Rover and secured shade netting to the ground. Then JB hiked with a full backpack to the cave, where he planned to keep watch for three nights to see rhinos as they came to drink. Soni and I waited at the vehicle. Each day Joel hiked back out to us and reported his observations. One night he saw a klipspringer, a small monogamous antelope, thick in the hindquarters with small pointed horns—another night, a band of mountain zebras. He asked anxiously about our nights.

"Nothing really. I heard a crow once at sunset. Sonja saw a lizard. That's it."

The canyons blazed. Ground temperatures climbed to nearly 150°F. We were all happy when we moved on.

The Ugab River has carved a wicked canyon. Subsidiary entrances and inaccessible tabletops spread for miles. The river itself is discernible from near Outjo, south of Etosha, and ends its long journey at the Skeleton Coast along the cold

Atlantic Ocean. Flood plains, thickets of razor-sharp reeds, and unbridled erosion made for slow travel. We shoveled, dug, and reshaped embankments. Namaqua sand grouse burst from bushes. A black eagle soared above. Signs of hyraxes, steenbok, and klipspringers littered the riverbeds and the canyons. Not far away, in 1987, a young rhino had been fed upon by three lions—the last time lions had visited the canyons—but whether they had killed it was unknown. The last elephants here were extirpated in the 1950s. Cape buffalo traveled the dry river systems in the 1800s. Now, the nearest buffalo are hundreds of kilometers away, at Waterberg Plateau Park, introduced by humans.

On our eleventh day out from Okaukuejo, deep in the Ugab's central canyon, the Land Rover ground to a halt. Dirt pressed firmly against the chassis. Three wheels touched the ground; the fourth spun idly. Joel grabbed the high-lift jack. I looked for rocks to stabilize it.

Not far away, water bubbled from a high water table, ran over metamorphosed rocks, and disappeared under soft crust. Water was also the source of much unrest. Gemsbok dug holes, "khoras," to reach the water deep in the sand. Bill Hamilton, a Californian zoologist, and his coworkers had discovered that male and female gemsbok fought for the priceless resource. Females usually lost, and more died than males, one of the few cases in which adult females are less abundant than males in a population. During an East African drought, more elephant females died than males. This pattern differs from the general one in mammals where more males than females die.

With the Land Rover finally free, our search for fresh rhino spoor continued. We made more temporary camps; the loading and unloading went on. JB scrambled up the Ugab's steep canyon walls, scanning patiently for hours, hoping and waiting for a rhino to pass below. Soni and I erected more shade netting and passed the days swatting mopane flies, coloring, and reading storybooks. When Joel called on the two-way radio, Sonja excitedly piped back, "Daaadee" and "bye." Joel relayed his view above the canyon. "The mountaintops are shrouded in morning fog." Or, more often, "The walls are incredible, Cunningham. You've got to see this. I'll stay with Sonja and you climb."

The coastal moisture and the inland heat clashed at 2,600 meters elevation. We always hoped that the coolness would drift back to the canyon bottoms, but it never did.

High on a precipice rested a chalk-white skull. Two elongated canines, discolored and sharp, revealed the sex of its deceased owner, a male baboon. Like most larger mammals, baboons are sexually dimorphic. Males are bigger, and often more ornamented, than females. Charles Darwin was the first scientist to suggest that the reason for such sex differences was that males competed for mates, mostly through combat. The females went to the victor. Competitive qualities are favored by natural selection and, over time, passed on to future generations. For male baboons, these might be longer canines. Among deer, larger antlers confer an advantage. At some point, though, an excessively large structure may be disadvantageous.

Darwin actually posed two explanations of why males often had characteris-

tics such as horns, antlers, or tusks. He suggested that such traits signaled qualities that females might choose. For mammals, these ideas have not been
subjected to exacting tests, but contemporary biologists are trying to decipher
the relationships between the sexes, including patterns of mating. It is now
known that a strong relationship exists between the degree of sexual dimorphism
and a species' mating system. Polygynous species tend to be organized into systems where males compete intensively for mates and are sexually dimorphic. In
contrast, although monogamy characterizes less than 5 percent of all the world's
mammals, monogamous species tend to be monomorphic, the sexes equivalent
in body size with neither males nor females having ornate structures. Black rhinos are monomorphic, white rhinos dimorphic. Baboons and gorillas are generally organized into societies where a single male lives with a troop of females. In
such species sexual dimorphism is the rule. Humans are sexually dimorphic too,
and in most so-called traditional societies, polygyny rather than monogamy is
the norm.

"Aoh, aoh, aoh, aoh" resounded up and down the canyon. Joel sat above us, nestled somewhere on the steep walls. He wasn't the source. My first thought was
baboons, no doubt because of the skull that JB had found. Then from the shade
of the Land Rover Sonja and I saw 11 baboons approaching us.

Joel saw the baboons, but from his vantage point, he didn't think we could see
them or even hear them. He watched nervously as a large male came closer to us.
Joel couldn't remember whether tales of dangerous males were fact or fiction, but
he recalled Stuart Whitmore's battle with a baboon in the 1960s movie *Sands of
the Kalahari.* It was 10:20 A.M., 40 minutes until he could warn us by radio. Two
hours later, he joined us in the sweltering canyon.

"What'd you think when the baboons closed in? Were you scared?"

"No, I pulled out the spotting scope for a better look. They were about 100
meters away. I thought it'd be fun to see them better. Why?"

Joel paused, shifting his brown eyes toward a rocky slope, then refocusing on
me. His expression shifted from concern to chagrin as it dawned on him that there
had been no threat. None at all. From above, it all had looked so different. *Deja
vu,* I thought, remembering the rhino in Aub Canyon and me driving frantically
down the gully to rescue him. I'd been sure he was in serious trouble.

"It all depends on your perspective," he said, apologizing once again for his
tirade back then.

More days passed. We left the Ugab, headed beyond Doros Crater, and saw places
where ancient stones used by past humans were clustered near water. We visited
petrified forests, remains from when the Namib wasn't desert, bounced, and
sweated, and still found no rhinos.

One day our problem dawned on us. It wasn't that we couldn't find rhinos, it
was that we expected to find them. We anticipated data. Perhaps we shouldn't be

applying rigorous standards of academia for our research here. Africa was different. So were rhinos. This study wasn't going to be easy, but we couldn't give up, not yet. We just needed to figure out the best way to find them.

We continued our journey through mountains that were pale and dusty, hoping that Archie would be where he had promised—Sesfontein.

10

Through the Eyes of a Poacher

[Joel]

The Zessfontein oasis . . . has lured me for years like a Saharan legend. Zess-fontein is a self-contained world, one of those sanctuaries I am always seeking on my travels. There beside the water from the six fountains live the people of a dying race. . . . They are on their way out; but at least they have chosen a tranquil and remote spot . . . well over 200 miles from Outjo, the nearest village. [They] have few possessions apart from their cattle; so that unless lions burst into the kraals, or elephants loot the gardens, they have few worries.

So wrote Lawrence Green in his classic *Lords of the Last Frontier*. "It is no easy journey. Zessfontein . . . is ringed by nameless mountain ranges. Men, on foot, pack-animals, and sometimes jeeps, find several ways in and out of this 'lost' world."

Now, 45 years later, Sesfontein has changed, but not much. The stately Old German Fort still stands, but now it shelters cattle, pigeons, and bats, and is little used by humans. Nearby, dung and sticks mixed with mud form beehive-shaped dwellings. Water flows from communal spigots. Soil and sand whip through the air, the result of decades of overgrazing and gale-force winds. Lions still prey on the unsuspecting, but only to the south. Ostriches, springbok, and elephants remain local residents.

Throughout much of the Kaokoveld, elephants are now extinct. To the north of Sesfontein, Thomas Mutati, a Herero headman, described what they were like some 70 years ago: "This is a bad place for elephants. Now they are away, but if you made your camp under that tree in September, they would kill you. I have seen them come through this village, between our houses, breaking down the walls and killing my people, 10 or 12 people, men, women, and children. A whole herd came in, and even the large fires could not stop them."

We progressed slowly, although the main route to Sesfontein had been graded. Whenever we exceeded 30 kilometers per hour, we ricocheted like a billiard ball on the road's washboard surface. The other side of the road beckoned; following the local practice, I'd search there for softer substrate, fewer bumps. Sometimes donkey carts and cattle materialized behind blind turns. Other times hot air blasted through our open windows, delivering powdery dust that turned hair and skin to dun. The incessant bouncing weakened brackets, and once again the refrigerator broke loose, slid forward, and shattered a trunk, a K-Mart special.

This day we were fortunate. We arrived well within the three-day period Archie and I had arranged for our rendezvous. Sesfontein has no phones and no mail service; not even buses travel there. If we were going to work with Archie, we'd have to factor in the extra days needed to get and return him.

This was our first opportunity to determine whether Archie was as good, or as bad, as people claimed. Many thought that his sentence for poaching rhinos, a felony, was a sham—four years' probation, an assignment that could be spent in association with a conservation project. Why not imprisonment, at least a heavy fine? Sam Nujoma, ex-freedom fighter and now the country's first president, declaimed about the need to reduce poaching. A new fine of up to $70,000 and 20 years in jail were now possible. The magistrate's decision that Archie serve probation under his former employer, Save the Rhino Trust, was laughable to many. Some white conservationists claimed it was a slap in the face of justice. Black Namibians, too, understood the need to maintain live rhinos, but they didn't understand the desire for imprisonment. Couldn't the system work without it? Locals, even though most had never seen a rhino, knew the animal's value. Rhinos were one of the species that tourists paid heavily to see. But the dollars brought by rich foreigners rarely made it back to local hands.

Small children, boys and girls, some naked, ran to greet our Land Rover. Adults peered from their rondavels, their three-legged kettles standing over smoldering ashes. Chickens, high-stepping past broken glass and other debris, moved aside to make way for the American strangers. We tried to ask people where we might find Archie Gawuseb, but most understood only Damara or Afrikaans.

"Gawuseb, Gawuseb," we'd say, trying to pronounce it properly.

Blank eyes stared. No one seemed to know him. Carol was right. We should have had his photo.

A barefoot teenager in a torn dress pointed up a sandy spoor. Then a man appeared and smiled. He jumped on our hood and squatted. He directed us back and forth around the dusty village for more than an hour. Finally, he jumped off and held his hand out. Two rands poorer, we continued our search.

Becoming desperate, we tried clusters of huts. Not understanding local practices and not wishing to unwittingly breach them, we stood next to the Land Rover and waited for someone to come out to us.

One hut had a garden; another had donkeys. A third had an injured parrot in a cage with an old woman crouching by a midday fire. As we waited, a young woman appeared, her bright blue dress fluttering in the breeze. Her name was Herero. This confused us greatly at first; Herero was the name of a group of people, like Owambo or Damara, not an individual, or so we thought. We were wrong. Herero was Archie's common-law wife, and she was a Damara. Names of individuals that seemed strange to us were perfectly normal to them, names like America, Himba, and Today. Archie's son was Somebody. His two daughters were more traditionally named: Maiyola and Fabiola.

Twelve children surrounded the Land Rover, smiling and chattering but skittish. Two of the most curious reached for Sonja's hair. She tensed, clinging tighter to Carol, who opened the door and stepped out into the hot sand. The kids laughed, and then ran away, still laughing, speaking a language we didn't understand. An old barefoot man with a black hat limped over, using a cane. A door creaked open, and a lowered head emerged, crowned by a blue Nike visor, my old hat.

"Hello, my friends," said Archie. Carol and I shook his hand.

"Why couldn't anyone tell us where you lived, Archie?"

"We drove back and forth asking about you; Sesfontein isn't that big," added Carol.

Archie was quiet, then he smiled broadly.

"Everyone knows me here as !Huie."

This was his Damara name, not the one we knew—Argelius Archie Gawuseb. That name, a traditional Christian name, had been insisted upon by the South Africans who had run Namibia prior to independence.

His father recalled trips to Outjo during that time, blistering journeys in donkey carts, when security police provided three-day passes for visits into town. His father had accepted rule by South Africans. Archie had not. During the twenty-year war he had supported the resistance, SWAPO (South West African People's Organization), now Namibia's elected party.

Archie packed quickly, handing his mattress up to me on the rack. I cabled it and his yellow chest between the high-lift jack and six jerry cans. Three tires, two water jugs, and three trunks were already secure. Other supplies that we'd bought for Archie—a pack, boots, binoculars, and a North Face tent—went inside. Carol talked with Herero. Neither understood the other, but both smiled a lot. Sonja tried running barefoot with the other children, but her naked soles hadn't been hardened by years of shoeless play. Her unkempt blonde hair fascinated the kids, who stared, touched it, and pulled it.

I approached one of the older children, maybe eight or nine years old, and extended my hand. He stared, puzzled. *What did this stranger want?* Old people gathered. I reached out again. This time he reciprocated. Slowly and deliberately, I turned his hand over, palm upward. Then I raised my hand above his and moved it down, gently guiding mine until it touched his. I reversed positions, placing my hand under his. I gave him "five."

Other kids moved closer, giggling nervously in a language full of clicks and smacks. Soon all of us exchanged "fives." Archie knew the sequence well, since our first rhino the month before.

One child didn't participate, a little girl with pierced ears. She was very small, perhaps two-thirds the size of Sonja. Mucus dribbled from her nose. It was Fabiola, Archie's youngest. She and Sonja were the same age, exactly: both were born on August 29, 1989. Cunningham had given birth in traditional American style, in a well-staffed hospital. An older first-time mother, she had needed extra help. On the other side of the world—at a farm named Humor owned by Archie's parents—Herero had given birth without fanfare: no nurses, no physicians, no "modern" facilities. Herero did what women have done for millennia, she relied on local knowledge and family support. The coincidence was extraordinary and reinforced our picture of Archie as a whole person, not just a tracker. Archie, like every other caring father, had children to feed, a life to live, a family to love.

Archie relished life in the bush. Unlike most youngsters who have been exposed to western ways and lured from traditional farm life, Archie had been in the urban jungles of South Africa but preferred the natural world, its beauty and danger. Farms in the remote veld are nothing like those of Europe or North America. Crudely sculptured in wild country, they have few fences and no paved roads, just cattle, sheep, and goats. Archie was rooted deeply in a simple system, one with economic poverty and spiritual wealth.

As a boy, he learned a skill as ancient as humans themselves—tracking. Youngsters cared for the family stock. Goats released from kraals made of acacia and barbed tangles were tended throughout the day as they searched for food. When animals became lost, young boys found them by following their spoor. The more astute the person, the better the tracker.

Impressions in wet sand, toe scuffs in clay, displaced pebbles, even bent stems sent messages. The marks of hooves of different sizes or shapes, feet with different numbers of toes, fur, pungent dung, urine, even saliva all revealed the passing of monkeys, lions, and giraffes. Interpreting sign could be difficult, complicated by unpredictable winds from different directions, the time of day, the temperature, and the substrate. How long since the animal had passed? How many were there? Did the same animal return, or was it another? Did he or she enter this canyon or that one? Was it running or trotting, hunting or just moving?

The art of tracking may well have led to the origin of science, wrote Louis Liebenberg, a South African living with Bushmen in the Kalahari. Successful early hunters must have been rewarded more than those who were poor at tracking. Hunters relying on sound deductive logic must have been better than those who didn't. The very transition from primitive to modern culture was no doubt aided by those who could interpret sign—they had knowledge of zoology and anatomy, ecology and animal behavior.

Archie spoke Damara, Herero, and Afrikaans, some English, and Owambo, five languages in all. Yet he had never graduated from high school. What we still needed to know was just how good Archie was. Could he interpret footprints in the sand? If not, we were in trouble. Our study of rhinos in the northern Namib might be over.

As the Boat drifted south from Sesfontein, Archie sat in the back seat, squinting toward the bleak mountains. He waved good-bye to Herero. He wouldn't see his family for a month. Hot wind continued its assault, raising plumes of dust from the cracked earth.

"Would you like a cold Coke or juice, Archie?"

"Coke is my favorite." The conversation ended.

Distant buttes faded as the dust storm swirled. We drove through canyons and riverbeds, places that had rhinos when Archie's father and grandfather were boys, places that have them no longer.

* * *

Protected areas, conservation enclaves, and game reserves are recent additions to the African continent. The oldest have existed for a hundred years. Wildlife conservation is a recent human endeavor, an experiment. In an increasingly crowded world, conservation is seen as good because it offers a chance to protect wild places, to maintain healthy ecosystems, and a chance to educate people. But many Africans have a different view.

Conservation areas have been designated in places where local people also lived, subsisting on the land—as pastoralists, hunters and gatherers, farmers, and traders. As colonial administrators imposed laws to protect wildlife, a system foreign to local people was instituted. Restricted access, relocation, and the prohibition of use followed.

Foreign ideology superseded human rights. Genocide was practiced, and land was taken. Huge reserves were born in the name of national interests. Masai were obliged to leave parts of the Great Rift Valley in Kenya in 1904; in the same year 10,000 Hereros were killed by Germans after protesting the invasion of their land in Namibia. None of the horror is unique. Catholics slaughtered Protestants in France during the 1570s. Between 1650 and 1800, Boers killed more than 10,000 San Bushmen. During the nineteenth century, thousands of Tasmanians were eradicated in Australia. Native Americans were destroyed by colonizers from France, Spain, and Britain, and by United States citizens. Ten million Jews, Gypsies, Poles, and Russians were annihilated during World War II.

Greed, killing, and the suppression of indigenous cultures has also destroyed noble species. On the northern prairies of the United States, white settlers and soldiers massacred bison, an important source of food once available to the Lakota (Sioux) Indians. After the Lakota were forced onto reserves, government rations failed. The natives, having already lost their cultural identities, forced to wait for government handouts, chose rebellion. Black Elk, said in 1875, "Once we were happy in our own country and we were seldom hungry, for the two-leggeds and the four-leggeds lived together like relatives, and there was plenty for them and for us. But the Wasichus [whites] came, and they made little islands for us and other little islands for the four-leggeds and always these islands are becoming smaller."

Henry Stanley reported from his 1887 travels in central Africa:

There is only one remedy for these wholesale devastations of African aborigines . . . seizing upon every tusk of ivory brought out, as there is not a single piece nowadays which has been gained lawfully. Every tusk, piece, and scrap in the possession of an Arab trader has been steeped and dyed in blood. Every pound weighed has cost the life of a man, woman, or child; for every five pounds a hut has been burned; for every two tusks a whole village has been destroyed; every twenty tusks have been obtained at the price of a district with all its people, villages, and plantations. It is simply incredible that, because ivory is required for ornaments or billiard games, the rich heart of Africa should be laid to waste.

Twelve years later, Joseph Conrad described a demoralizing journey up the Congo River to a scene of unwavering exploitation in *Heart of Darkness*.

Such tragedy is ubiquitous. In Namibia, Etosha National Park was established as a massive 99,000-square-kilometer reserve in 1907; in 1970 it was reduced to its present size of 22,700 square kilometers. Historically, the park stretched to the cold Atlantic coast, where its fauna included more than savanna-dwelling lions, elephants, and cheetahs. Southern elephant seals, pilot whales, green turtles sometimes a meter long, and wayfaring jackass penguins might have graced its shoreline. But native people experienced relocation, segregation, the denigration of traditional lifestyles. Spirits were broken.

Namibia was a protectorate of South Africa under a treaty signed by the League of Nations. The mandate was changed by the United Nations in the 1970s, and South Africa's continued occupation was made "illegal." The new Namibian government inherited a host of environmental and land use problems. In the north, people were maimed by mines left ungathered after the war. In 1990 13 people died from these—22 in 1991. Countless amputees survive.

Other issues were deeply rooted in the archaic policies associated with "homelands" and apartheid; 50 percent of the usable land was owned by 2 percent of the population. Some Damaras were uprooted and stuck in barren wastelands, deserts that presently house the continent's last unfenced black rhinos. Now, however, it is the poor people who have a chance to benefit from the rich tourists who come to see the rhinos and other wildlife. Despite unemployment rates of 80 to 90 percent in and around Khorixas, those living on the edge of the Namib may yet have jobs and a chance to exercise choices about the management of their natural resources.

Changing attitudes, the possibilities for reempowerment, and committed nongovernmental organizations (NGOs) are all helping to reshape Namibia's conservation infrastructure, especially in the Kaokoveld. NGOs ranging from the Endangered Wildlife Trust and Southern African Nature Foundation in South Africa to Save the Rhino Trust have supported personnel, bought equipment, and hired locals. The Ministry of Wildlife, Conservation, and Tourism (now the Ministry of Environment and Tourism) has research, management, and planning divisions working with these groups. The efforts of Garth Owen-Smith and Margie Jacobsohn have been key. They were among the first to hire ex-poachers as game guards, most appointed by local headmen—although similar practices are being used in Zambia, Zimbabwe, and elsewhere. It was Garth who said, "I believe the local people hold the key to survival of wildlife in Africa. Unless they see some material benefit from conservation, they're not going to support it when the chips are down—when there's population pressure and pressure from grazing."

Whether humans can solve these problems remains to be seen, but attempts to include local people in the decision-making process represents a new era in conservation.

*   *   *

Silver-leaved catophractes, brushy and reaching over three meters tall, ruffled in the breeze. "Orrr whip, orrr whip, orrr whip." The early morning tranquility was broken by the sounds of an agitated korhaan. Wings beat furiously, irate hornbills squawked. "Qua ha, qua ha, qua ha," barked zebras. Hooves pounded. Something was wrong.

Bursts of gunfire followed. A zebra lay dying. This was far away from the well-traveled tourist areas. Four men stood by with .303 rifles. Dave Murray, the blond, bearded leader of Etosha's Anti-poaching Unit (APU) approached with the other members of the APU team. Two men dropped their weapons, two others fled. At the crime scene the remaining poachers continued eating the splayed zebra, consuming it raw. Knowing that they were starving, Dave allowed it. At the jail in nearby Kamanjab, the men were released. The jail had no food to house them; they would just have to come back on their own for a hearing when the judge next visits.

Another episode: an errant springbok wanders near Okaukuejo's service employee housing section, known locally as "the location." It's where laborers live, away from the clean bungalows, tourist quarters, and swimming pools. It's an area hidden from the visitors, where water comes from the familiar communal spigot, relatives are crowded into one-room shacks, and malaria-bearing mosquitoes find easy prey. It's a place of poverty and unrest, of alcohol and little food. It is where our friends from the restaurant, Hans and Fernando, live. It is also a place where we are uncomfortable. Our presence invites stares.

But the springbok is unafraid, habituated to park visitors. It is also dangerously close to hungry people. A young man with good aim seizes a rock the size of an orange. At astonishing speed it sails to the unsuspecting springbok, and the crack of stone against bone reverberates. More missiles are hurled. Sticks whack the prone animal. Death is swift, the carnal scene played out within 500 meters of Okaukuejo—within one of southern Africa's premier national parks.

On a weekend earlier in the year, more than 150 snares—cruel, painful, and deadly—were found, all adjacent to "the location." They had been strategically placed on paths that led to Okaukuejo's floodlit water hole, the only dry-season water for miles and miles. Rusty wire wrapped around tree trunks was designed to kill—to cut and tear with each pull—naive prey, animals that had long lost their fear of human smells. Razor-sharp metal from the tops of 55 gallon drums, carefully honed, was placed on the ground. Death waited for unwary feet. These were traps of anguish, set not by people seeking gall bladders or horns, not by people marketing products. The people here simply hoped to add protein to their diet. They were the unemployed, or people who, when working, would be lucky to make $100 per month, people with families. It didn't matter whether they lived in a national park with sacred wildlife or on the fringe of desert, or even in a faraway city. They were people trying to survive.

In Tanzania, 60 percent of the killing of wildlife is illegal, primarily subsistence hunting to feed hungry mouths. In the Serengeti some 200,000 ungulates are taken illegally each year. To the south, in Mozambique, the poorest country on earth, the average person makes a mere $80 each year. Most people have no income.

The fact that people are famished does not justify breaking laws, but it is far easier to understand that type of motivation than large-scale commercial poaching operations, often run through organized gangs. A barter economy, including the trade and sale of wildlife products, has gone on for centuries. Today's story is different. Modern weapons coupled with the lure of money from foreign markets drive unscrupulous traders to exploit impoverished people. The result has been the decimation of animal populations. Ostrich feathers were once fancied as ornaments in Europe and the Americas. Gall bladders and tiger bones are used medicinally in Asia. The illicit trade in neotropical parrots continues to annihilate populations.

Civil strife and war have contributed to the demise of rhinos. Among early casualties were black rhinos in Nigeria, gone by 1945, and northern white rhinos. Sudanese rebels invaded Zaire's Garamba National Park, and the number of "whites" dropped from an estimated 1,200 in 1963 to less than 50 five years later. The story is similar throughout the subcontinent, although the causes vary geographically. Angola's civil war lasted nearly 20 years. Few conservationists have braved conditions there to determine the biological effects of war, but it would be surprising if more than a very few rhinos were left. No one really knows. Parallel stories are depressingly common. A 1981 report by Markus Borner in the British journal *Oryx* detailed the "Black Rhino Disaster in Tanzania"; another, by Eric Edroma, declared, "White Rhino Extinct in Uganda." Black rhinos are virtually gone in Zambia; a report in the same journal carried the subtitle "Poaching Pays in Luangwa Valley, Zambia." As of 1980, Ethiopia was estimated to have 20 blacks rhinos—10 in 1984. In the Central African Republic's Bamingui-Bangoran National Park, black rhinos persisted in the mid-1980s until Sudanese horsemen began their relentless search for elephant ivory there. By 1966 less than 50 black rhinos remained in Somalia. The species is probably gone from Sudan, as well as Chad, Zaire, and Rwanda. In Kenya, 75 percent of Amboseli's national park rhino deaths resulted from spearing, reprisals by Masai warriors in modern land disputes. Aerial surveys in Botswana in 1992 found no black rhinos. South African, Kenyan, and Zimbabwean rhinos in well-protected reserves are reproducing. Those in the Namib Desert and Etosha hang on and increase, despite significant setbacks due to poaching in the last decade.

Why poachers poach is easy to understand: money. Africa remains a troubled continent. Of the world's 36 poorest countries, Africa has 29. Literacy rates are low, in Namibia about 30 percent. Infant mortality remains the highest on earth, and 90 percent of Africans live in poverty. Joblessness reigns. Civil strife has affected nearly every country in the last 15 years. Money spent on conservation by governments is money not spent directly on Africans. In westernized countries, it is easy to talk of solutions. Yet to unemployed and often undernourished people, wildlife products often provide substantial gains. Poaching makes sense. The fact that Zambians travel to Zimbabwe, Sudanese horsemen to Zaire and the Central African Republic, and Somalia shifta to Kenya indicates that poachers do not have to be locals. All they need is to be risk-takers, and to be offered incentives. How else can the killing of three guards protecting rhinos in Kenya's Shaba Game Reserve be explained? By 1993 armed guards stood sentry 24 hours a day over rhinos reintroduced in Malawi's Liwonde National Park.

How much poachers earn for each rhino is difficult to say. Cash rewards vary geographically, according to supply, demand, and local economics. The per-horn prices appear to range from $100 to $360. Given that the risk of detection is low, the chances of capture and imprisonment lower, and—despite the deaths of more than 150 poachers in shootouts with paramilitary Zimbabweans—the possibility of death remains low, it is not difficult to understand why unemployed people might poach for more than meat.

*   *   *

Common terns and cormorants danced above the Atlantic's angry whitecaps. Sun cut through the early morning fog, revealing dew and grass. Several years earlier, a rhino had walked to within three kilometers of the coast. We wandered, exploring the sandy hummocks.

The tracks of a lone brown hyena led to last night's meal, a Cape fur seal, its carcass rotting near one of the many shipwrecks characteristic of the Skeleton Coast. The scattered bones of a baleen whale littered the dunes where black-backed jackals had made their middens. But lions, once the most awesome predator along this desolate coast, are gone. They ate pilot whales and jackals, and were known to drag fur seals more than two kilometers.

They even captured white-breasted cormorants. But the coastal lions that roamed here in the 1980s are now extinct—molestation of stock was the charge, death the sentence.

The purpose of our coastal side trip was to find the skull of a female rhino that had died earlier in the year. We needed to bolster our sample, to help us assess what, if any, relationships exist between horn size and age, head size, and sex. Although rhino skulls are plentiful throughout the continent, many of their faces have been hacked off, making it impossible to know whether horn size was influenced by age or sex. And in the comparatively few cases where skulls and their horns had been measured, the sexes were usually unknown, since the heads of males and females are the same size. This lack of conspicuous sex differences was a problem. It wasn't possible to determine the sexes of poached animals or even of live populations. This skull was unusually valuable: we knew our elusive victim was a female, because she had been discovered only a few days after death.

Archie knew the guards at the gated entrance to Skeleton Coast Park. They had pointed toward a desolate valley surrounded by sterile mountains. The sole sign of humanity was a trash heap. Inside a rusted steel bin rested a leathery mass of dried flesh and bone.

"Rr-ino skull." Sonja's speech was improving.

I measured the zygomatic width, occipital height, and facial breadth. We all struggled to separate the jaws to examine the teeth for aging. This female had been old, more than 30 years of age.

We wondered about the other skeletons we'd found. One, near Doros Crater, had died in a fight with a bull called *Mad Max. Max* had horns more than half a meter long, 20 centimeters across the base. He roamed a vast area, some 2,000 square kilometers. Evidence near the carcass, crushed vegetation and disturbed soil, indicated that a battle had occurred. The recovery of horns had suggested that poachers were not involved. We still needed this male's skull for measurements; it would be another valuable data point for males. It was somewhere in Khorixas. The partially burned carcass of what we now knew was a young rhino, the one we had found below the Bushmen paintings at San Cave, remained puzzling. Oddly, no one seemed to know about it. If it had been killed by poachers, why was the skull gone? Why was the body burned? Had humans taken it? Like us, Archie too was in the dark, although he had worked the Doros Crater area for years.

Sonja's *Jack and the Beanstalk* tape ended. It was Archie's turn. He picked his favorite, Sergio Leone, but was uncharacteristically talkative, describing the day he was gored.

"I had been following a rhino and found two. When I was taking their photographs, *Stumpy* [a third rhino] knocked me down. Then, he was standing over me and I felt blood dripping down my back. It was warm. My arm hurt. I thought I would die."

Archie turned to me and started laughing. He hadn't been badly injured. Most of the liquid wasn't blood. It was saliva dripping from the rhino. He showed us the scar on his underarm from the attack.

"I am not afraid of rhinos, Joel. . . . Tracking alone is dangerous in rivers when I cannot see, but rhinos do not scare me. I like rhinos. I like watching rhinos."

"What about elephants? Do you think they're dangerous? What animals are you afraid of?" I grinned, goading him.

"I am not afraid of snakes or scorpions. Lions are usually afraid of people. At night, when I am in a tent, I do not worry about them."

"Yes, but Garth was bitten by a lion," I interrupted.

Ignoring me, he continued, "Elephants are a problem, a big problem. People in Sesfontein are afraid of elephants. I like to see elephants, but not too close."

We moved inland from the coast, climbing up over more mountains and squeezing through narrow passes. Remnants of humans appeared and disappeared, a thatched structure here, a clay dwelling there. Without rains, this land was no good. No good to people, that is, people with cattle and goats. A gemsbok punctuated the dreariness of endless mountains.

We entered a valley of petroglyphs and pictographs, a place called Twyfelfontein—"doubtful fountain." Thousands of etchings, some perhaps 2,000 years old, decorated massive sandstone. Weathered paintings faded into stately rock. Shallowly carved, red-stained lions with long tails stood next to giraffes. Elephants loomed above zebras. Cunningham and I picked a careful path, climbing boulders that were pinched into tight passages. We traded carrying Sonja. Delicate human figures were hidden under overhangs; chiseled rhinos adorned arches. Anvils of light filled the valley below. Although we had seen petroglyphs and pictographs before, neither of us was prepared for the sheer volume of exquisite art here.

A dark, distant silhouette appeared at the Land Rover below. Reality returned. Archie was waiting.

We plowed forward, anxious to attack the Ugab with a tracker. We passed more signs of humans, not those of modern people but of ancient nomads moving back and forth between high mountains and coastal deserts. Encampments, stock posts and rock shelters, beads made from ostrich shells, even primitive digging tools and stone pestles were in the Ugab, its dry tributaries, and on the mountain to the south.

We visited Matheus, one of two Save the Rhino Trust employees at a semipermanent camp pinched between the Ugab's canyon walls. Neither had been able to identify rhinos, for they had no binoculars.

We never knew whether to leave food or beer. How could we not leave at least a few supplies? But this was not America. Arrangements were already in place. Namibian scouts worked for a few dollars and little more. Mealie was delivered—we didn't know what else. They did without amenities because there was no choice. Matheus had a wife and a baby; they were making a go of it in the remote heart of Ugab's furnace. We drove off. As we pulled away, tins of fruit and beans that were hidden under the Land Rover emerged. In the rear-view mirror, we could see smiles and waves; the supplies had been discovered. A cold six-pack of Tafel beer lay in their midst.

Three days of tracking uncovered spoor but no rhinos. Near sunset on the fourth day we glimpsed three rhinos moving quickly down a deep canyon. Immediately Archie leaped into pursuit. I struggled to keep up. Soon it would be too dim for photos. I pushed harder.

Strong erratic winds whipped from behind us. The rhinos smelled us, whirled, and galloped. We'd lost them. If I'd been quicker before the winds exploded, the day would have ended differently. Growing old is difficult enough, but now my ego was bruised. Archie was in better condition, much better. We both knew it.

Another time we drove 220 grueling, rock-filled kilometers over three days without seeing any fresh rhino spoor. But persistence had its benefits. We now knew which areas didn't have rhinos.

Our luck changed. A young adult rhino with a huge tear in his left ear showed up in a narrow canyon under the only shade. He had been dehorned. He also smelled us before I was close enough for the photo and fled in panic, but we found him less than half a kilometer away. This time we got photos, both profile and frontal at close range.

The next day we were checking other areas far to the west of Doros Crater. As we dropped into a canyon I downshifted, pumping the clutch. Nothing happened. I pumped again. No leverage, no tension. The gearshift wouldn't enter first gear. It wouldn't go into second or into reverse.

We had no clutch. Base camp was 60 kilometers away, Khorixas 200. We looked into the back of the Land Rover. One spare trunk with tins of food in it sat next to the refrigerator, which held five pint-sized boxes of juice. Another case of juice sat in the back next to 20 liters of water. At least we had enough for a few days. Archie and I struggled to find the problem. Sun pounded the canyon's walls, and we dripped sweat. Helpless, greasy, and thirsty, we squatted in the scant shade and discussed options.

We couldn't shift into gear with the engine running. The car would start in gear but lurched forward like a runaway train. We could stop its forward momentum by yanking the gear into neutral, but then we were frozen—a catch-22. Even if the gear would engage, the car couldn't turn in the space available, at least not in the direction we needed to go. Speed-shifting was out. Our retreat was blocked by knee-high boulders.

"Maybe we can drive over them," Carol suggested.

It was an idea worth considering. If the Land Rover navigated the boulders and didn't roll, then the only problem would be getting from first to second gear.

"If this works, we won't be able to stop. Lean in the opposite direction to balance. Load Sonja and the toys, and let's get out of here."

Twenty minutes passed. We cleared the small rocks and piled dirt in front of those too heavy to move. With the Land Rover in gear I turned the ignition. We staggered, then chugged, hit the first rock instantly, reeled and smashed the bumper, swayed to the side, and then climbed up. Rocks disappeared below us. The Land Rover had a mind of its own; I only had to steer. Deep sand lay ahead. Momentum was critical. I accelerated, switched the ignition off, and forced the gearshift into second. Then I jump-started the engine, slamming my foot back onto the gas pedal. We hit sand at 30 kilometers per hour and drove on, with broad smiles worn by all. When the substrate became firm, I managed to force the gearshift into third. The sun sank, turning Doros Crater's black rocks to fiery pink. Base camp appeared. Khorixas was a mere 140 kilometers further.

At Khorixas, we desperately hoped for a good mechanic. But no one there could fix our problem. The master cylinder was a sealed unit in new models like ours. We'd have to go to Windhoek. Driving clutchless in traffic, switching the ignition on and off, was not an inspiring prospect, especially in a Land Rover filled with gear.

Before leaving we let Blythe know what had happened. We also told her of the new rhinos that we'd found. She thanked us, remarking that the photos we mailed earlier would be good for reidentifying individuals. Then came rhino news. Blythe and ministry officials had known all along of the charred remains near San Cave. The young animal had died from anthrax, and its body had been burned to minimize the spread of the deadly disease.

Cunningham and I were baffled as to why no one had told us. We mused about the still-elusive dehorning records and photo files, and we jokingly suggested to each other that they must have been burned with the rhino. We were in the dark.

Archie opted to stay with friends in Khorixas while we drove on to Windhoek for repairs. Before we left we met some enthusiastic Peace Corps volunteers teaching science and English at the local school. They were completely delighted to be in Namibia. We reminded ourselves that, despite our problems, we were lucky to be in Africa, and to be studying wild rhinos.

In Windhoek, with the Land Rover in a shop, we learned that Dan Quayle, the vice president of the United States, would be at Etosha in 48 hours. We were among a few conservationists invited to meet with him there. We looked at each other, shaking our heads. What an honor. But how could we possibly go back to Etosha now? We still had no clutch. We had just arrived in Windhoek from the field and were covered in grime. Quayle's visit had apparently been planned for months, but we'd just been told of it. Even if we rode there with someone else, we were exhausted and stressed and felt sorry for Sonja. Adding two more days to the three she'd already spent in a vehicle was too much to ask of a two-year-old. We hoped people didn't consider us ungrateful when we declined the invitation. Later that night, we recognized the vice president's bodyguards at an Italian restaurant in town. Loud and aggressive, they offered U.S. quarters worthless in

Namibia, to the waitresses, bragging that American money was worth 10 times the rand.

It was September 13, Cunningham's birthday. We celebrated by staying at the Safari Hotel. Not only was there a television, there was CNN—news from America. The hotel also had free video movies chosen by the local desk crew. Anxiously we screened the day's selection, hoping for good ones. We raced through the list: *Mirage, Beast of War, Damned River, White Dog—Trained to Kill, Friday the 13th.* It was Friday the thirteenth, we were doomed.

By the time we checked out, neither the videos nor the phone were working, despite the hotel's self-professed four-star status.

A new sealed master cylinder had been flown in from the coast, and the Land Rover was drivable, but our return to Khorixas and Archie was delayed slightly. The bank had exhausted their supply of checks and suggested that we carry cash.

"For the next three months?" exclaimed Carol in disbelief. The bank reluctantly agreed to mail us new checks at Etosha. It had been a bad few weeks, a comedy of errors.

When we arrived in Khorixas, Archie was there, waiting. We liked Archie. He was dependable, he worked hard, and he seemed to enjoy us. Our lives were intertwined. I knew his thoughts and signals. He knew mine. I read his eyes, he read mine. A bond was developing.

Twenty kilometers out, the master cylinder seemed to explode. Again we couldn't get out of neutral. It was the same problem a week later. This time we jimmied the gearshift and ignition, limped to Outjo, and got the car fixed at Weiman's Garage. The next day we were back in Khorixas for Archie.

Poor Archie. Not only was he forced to wait and wait and wait for us, but his other problems were serious. He had been told by Blythe not to return to Sesfontein because he was supposed to be doing continuous community service. Blythe had arranged for Archie to report to Hobatere, an elegant tourist lodge set among the hills and kopjes west of Etosha. He was to work the fence line, finding and repairing the holes punched by elephants, warthogs, or people. Although Blythe's promised supply of mealie had not materialized, the lodge would offer food, even an occasional mountain zebra steak. There was also a building with a stone floor for him to stay in, with a communal spigot was outside.

If Archie did a good job mending fences, there was a possibility of being hired, perhaps as a bartender, or to lead game tours for foreign visitors in Hobatere's open jeep. It was an opportunity shared by few Africans. But despite the safety and options that Hobatere offered, Archie considered it a sentence akin to prison. Not only would he have to work for free, he couldn't see his family.

The desert had been a disaster for data and expensive in time and money. The moon was waxing. It was nearly time for the next round of nocturnal observations in Etosha. Rhinos would be at Zebra Pan. We vacillated. Should we take Archie to Hobatere now or give the desert yet another chance, forgoing data opportunities at Etosha until the following month? Neither Carol nor I wanted to make the decision to write off the desert.

We recalled the words of Pete Morkel, Etosha's talented, energetic veterinarian: "Someone might say that you complain, when what you need to do is get off your butts and go out and find rhinos."

He was right. We needed to go to the desert and find rhinos. Jerry cans, water jugs, and food were stuffed onto and into the Boat. Nurturing a shred of hope, we pushed off for the Namib.

A routine developed. Carol sat in the front, Sonja remained on the box between us, and Archie crowded into the equipment-stuffed back seat. But when we left the main road and entered the desert's realm, Archie sat in front or up on the roof rack. At times Sonja rode up there cradled in his arms, enjoying the rushing air and the veld.

From above, Archie was in control.

"Turn. Stop. Left. Gemsbok. Giraffe." He cued us to animals, skeletons, and other signs of life or death. I was the driver, focused on sand, faint spoor, or rock. Like a goshawk, Archie spotted distant animals—a kudu under a tree; an aardwolf, its striped coat camouflaged in the shade. He told us what he saw, which side of the Land Rover it was on. I noted it all. Although our rhino data were meager, we were getting information on the abundance of other desert animals.

Dung, piquant and fibrous, lay in the path. Archie's boot pushed squarely into its middle, splitting the dry shell and sending vibrant molecules into the air.

We knew the plan: fill water bottles, synchronize watches, and grab the two-way radios, the big lens, and the monopod. Get the rangefinder and the two cameras. Take lunch.

We'd try to be in touch with Carol at the top of every hour, but at best our radios would transmit only five kilometers. She would stay with the car and move hourly to attempt radio contact. Once we had a few desert rhinos under our belt, I promised, she'd track and I'd babysit.

Archie led across the gravel plain, where dusty footprints reflected the morning light. We passed spiky bushes with fine gray leaves and bottlebushes with thick stems. The half-chewed remains of a giant euphorbia stalk, dripping with milky latex, pointed the way. We hoisted ourselves up and over the next ridge. Large tracks with immense patties of dung lay along our path, footprints three to four days old—a small band of elephants with at least two babies. I remembered Archie's fear and Garth's warning about elephants as our steady pace continued.

Two hours passed. My throat was parched, and my body stank. We entered a huge valley. Archie pushed toward a lonely tree, the only spot with shade. A mine field of different-sized dung piles lay under the branches of the slender white boscia. Gemsbok, kudus, springbok, even rhinos had found this ephemeral refuge. So had ticks. Locals believed they were attracted by motion; it was more likely that the carbon dioxide exhaled by resting mammals attracted the ticks. Archie watched the horizon from under the shady branches. I moved out of the realm of the ticks and sat in the sun.

Moments later Archie was on the move. I followed, agonized by each step; my ankles twisted and turned, my toes rubbed against the hard leather. Upturned pebbles signaled the rhino had also been this way. Then, nothing. *How could a 1,000-kilogram rhinoceros disappear?*

"Wait here." Archie was gone.

I sat alone, daydreaming and swatting flies. A lappet-faced vulture drifted by, its crimson head and white trousers lit by the sun. *Has something died?* I slid my radio out of its case. It was 10:59, almost time to call Carol. We had missed the 10 A.M. call.

"Drive seven to eight kilometers to the east. You'll see a faint two-track as you approach the narrow canyon. We're near the triangle-shaped peak, and to the west of a charcoal-colored mountain. Wait there, can you read me?"

"I'm not exactly sure where you mean, but I hear you fine. I'll find you. Wait, someone wants to say hello."

"Good bye, Daddiii."

Archie and I reunited in the next valley. Like a bloodhound, he was already back on the fresh spoor. Wind blew. He angled a shallow kick at the ground and watched dirt explode from the parched surface. Its trail revealed the wind's direction.

Archie was in control, intense, staring toward the shadows cast by every bush. *The rhino must be close.* We rechecked the wind's direction, making sure the rhino wouldn't smell us. Here there was no escape. No trees or car, even the bushes didn't have sturdy branches. Just us, the bare rock, and a half-dozen pliable euphorbias.

We stepped and waited. Then we stepped and waited again—and again. Each time we repeated the procedure, we peered around a new bush and moved forward a few steps. We still didn't know whether a rhino was even there, but we well appreciated the deadly consequences of a mistake.

Archie had extraordinary senses. He could probably track an ant across a boulder field. I remembered how, 25 meters from a hidden rhino, he had heard a quiet flapping sound—the rotation of *Mad Max*'s torn ear. Now, we both struggled to detect the rhino before it detected us.

Silently we removed our packs and placed them on a slight ridge. The first three bushes we looked in were empty. But if the rhino was nearby, I'd be ready. The 500-millimeter lens balanced on the monopod, the black-and-white film was in the camera, and the rangefinder was strapped around my chest. Archie carried the smaller lens and a camera with color film.

We moved closer and knelt in the open, as vulnerable to attack as a springbok fastening its eyes on an exploding cheetah. Eight minutes passed. We struggled forward, each step crunching a thousand tiny stones. At 100 meters a euphorbia blocked my view. Archie lifted his binoculars. He stiffened and jerked backwards. I recognized his body language—rhino. My pulse increased.

Gently I stepped to the spot where Archie was standing. The rangefinder indicated 78 meters. Silently I closed the distance, while Archie moved to a safe spot and readied himself to enjoy the show.

The rhino had been asleep. Its ears whirled around. It no longer slept, but it hadn't yet stood. It had been dehorned, and now 12 centimeters of front horn and 6 centimeters of rear horn covered flattened stubs. At 65 meters, I squeezed the trigger.

Click.

Wind muffled the sound. I waited for my next shot, not thinking about escape. I was too excited about getting the photographs. I hoped the rhino would soon stand and face us. Seconds ticked by, ears rotated. I looked at Archie helplessly, wishing he had an idea. We stared at each other. This was a standoff between the rhino and me. Archie could do no more.

I still needed a facial shot to confirm the rhino's identity. The binoculars clung to my side, but I was afraid to lift them, thinking the noise or motion would seal my fate. Already I was too close. I looked again through the telephoto lens. This time the view was different. At the top of the left ear was a tiny tear. *Is this enough for identification?* No, I needed a facial photo. I grabbed a golf ball–sized rock and debated whether to throw it. If the rhino stood up, I could probably get the second photo. I couldn't wait any longer and hurled a fast ball, purposefully bouncing it in the dirt in front of the resting animal.

Before it connected, though, the rhino was up and charging hard toward me— head down, horns low. Only 50 meters separated us, and the gap was closing. I dropped the camera, lowered my head, and pumped every muscle toward Archie, trying to overtake him, hoping to pass him, anything but to be last.

This can't really be happening. Not to me. No.

The rhino gained. Less than 30 meters separated us. Jagged, uneven rocks lay on my right, smoother ones on the left. I turned right; the animal was only a few body lengths away. Terrorized, I maintained my panicked path, bracing for the inevitable.

A long second passed. The rhino veered from the rock clutter. Archie smiled. I didn't, at first, but soon the image of the ridiculous scene had me chuckling.

"Bet you never saw a white man run so fast."

Archie didn't understand American humor.

We talked about the charge. Millions of human scent molecules must have exploded off the rock, leading the rhino directly to me. I raised my hand, Archie lowered his, and we connected loudly, over and over, in ecstasy.

Carol came on the radio at 2 P.M.

"Where are you? Any luck? I've called each of the last two hours."

"Cunningham, we got the rhino and a photo. It was *Mystique*."

"Yeahhh."

Days later Carol tracked with Archie. My feet were blistered and bloody, yet we had been lucky, eight rhinos in five days. But now, a wound caused by an acacia thorn in my foot had swelled badly. I needed a break. Not Archie.

Instead of tracking on gravel plains, he and Cunningham toiled in the dense brush of a dry river for more than four hours in 100°F heat. The trade-offs were obvious. Because of the crackling vegetation that littered the ground, rhinos could hear humans before being seen and move off. Also elephants loved river bottoms; their tracks and immense dung were becoming more common daily.

"Open places are good. I like it when I can see ostriches at five kilometers. The riverbeds are too thick," said Archie.

On the positive side, trees were available there for climbing. I wondered whether Carol really didn't know how to climb. I remembered her petrified face one day in Wyoming. In grizzly bear country, she had admitted that she didn't know how. She had never needed to learn.

At 5 P.M. Sonja and I talked on the radio to Carol.

"The rhino bolted from the river. We lost the spoor and are heading back. Drive four kilometers west."

Soni and I rocked along for 45 minutes. Two ant-sized bipeds appeared in my binoculars. Dwarfed against a red hillside, Carol and Archie strolled, shimmering in the heat in the majestic, rhinoless, valley.

It would be well after sunset when we arrived at camp. Two hours of driving remained but only one hour of light. Again, Sonja would have dinner from a tin. At least she received vitamin supplements. Archie still rode lookout above us. Carol and I talked.

"Were you afraid, tracking with Archie?"

"No, not really. I want to practice more with the rangefinder. I'm not sure I've got it."

We confided in each other. We were pleased with ourselves—20 percent of the dehorned rhinos had been photographed. Soon we'd be able to calculate rates of horn regrowth.

Fists pounded frantically on the roof.

"Rhino, rhino," screamed Archie.

There, crossing a gravel plain into the fading sun, was the rhino, seven kilometers from where Archie and Carol had lost its spoor. If we wanted this last chance before dark, we'd have to move quickly. A lone tree with thin branches stood on the naked plain. The rhino was not alarmed, just walking—quickly. It was a female. The wind blew stiffly from behind her. Knee-high boulders blocked the Land Rover. We'd have to walk. Carol volunteered to go with Archie.

"Do you want a weapon? It's very open."

"No—it's okay," said Archie.

"Do it, you'll feel better," I said to Carol. "Just in case. The noise might scare the rhino."

Cunningham strapped the .44 magnum to her waist. The two were off, determined, plucking their way through the rock-strewn landscape. Soni and I sat talking, shielded from the wind.

"Read to me," she said, looking longingly toward her book, *The Little Engine That Could*. Although worried about Carol, I began to read. On page 6 my attention was shattered. A gunshot. My mind raced. *She shot her foot.* Only one speck was near the tree. I looked through my telephoto lens, the only optical equipment within reach. Dirt trailed in gusts behind the rhino as it ran toward the tree.

Later, Carol wrote in our journal:

> I walked as fast as I could, holding the rangefinder and camera, the holster whacking into my hip. The rhino walked parallel and fast. Because she saw us and picked up her pace, we ran the last 15 meters to the tree. I stood behind the tree's thin trunk as Archie tried to climb. He slipped, hung by one arm, and tried again. He slipped again. I thought, *How will I ever get up?* I held the camera, and he struggled but finally made it. The rhino kept coming. I asked Archie, "Should I shoot?"
>
> "Yes, shoot the gun, shoot the gun."
>
> I fired, to the right. She didn't slow. Her head was down, still coming. My ears filled with the noise.

I sat with Sonja, paralyzed, watching. Another gunshot pierced the steady wind.

> The rhino wasn't responding. I passed the gun into Archie's lowered palm. He reached down with his other hand and circled my wrist. My boot bottoms kept slipping on the smooth trunk. I tried to find a notch but couldn't. I hung in midair. Somehow I managed some traction, the rhino still coming. Finally I was on the branches.
>
> Then I noticed small pains on my legs. Ants, hundreds of them, were on me, biting. I was breathing hard, Archie too. He said something. His lips moved. The gun blasts had hurt my ears, so I couldn't hear. I had no idea what he was saying.
>
> When the rhino got to the tree, she bolted 20 meters and then came back. This time, it was fine, enjoyable. I was safe up in the tree.
>
> Finally she ran off. JB came up while I was still trying to climb down. I'm happy, Archie's happy, Joel's happy. I'm glad, too, that the rhino is gone, and that it's over. And that we got the photos. We measured the seven meter distance with the rangefinder and laughed. Back at the Land Rover we examined our wounds. A branch gouged Archie's leg; blood dripped. There are raw patches of skin on my underarms and legs.

We named this female *Kokerboom* after a local type of tree.

We reached camp at 9 P.M., on a day that had begun at 5:50 A.M., a long but good day. I organized data and equipment, and we talked of tomorrow's plan: We'd return to the river two kilometers north of that gravel plain. Our thoughts weren't on *Kokerboom*. It was *Stumpy* we wanted.

Earlier, in the fading light, a rhino had crossed in front of us. Archie saw a sausaged-shaped stump of a tail. *Stumpy*. There were reports of at least two *Stumpys* in the Kaokoveld, maybe three. Two were definitely male. Tomorrow's tracking would begin some 29 kilometers away.

The dreaded east winds gathered force with the early light. They drew heat while steamrolling their way westward from the Kalahari Desert across the northern Namib to the Atlantic. Henna Martin described them in his 1940's book *The*

Sheltering Desert, commenting that for months prior to the rains they blew savagely. Now it was our turn to experience the 103°F heat at 10:30 in the morning.

Carol and Archie tracked in a windy river bottom where the temperature reached 109°F. Sonja and I waited, passing the time with drawings on the solar-powered computer. At 2 P.M. the intrepid trackers returned with splendid news. They had photographed *Stumpy* from a tree at 64 meters.

During the last nine days we had seen 13 rhinos, more than we had imagined possible. We had not been as successful as Garth, but it was a great deal more than one rhino per week. We were happy.

At another makeshift camp, we heard spotted hyenas, a lone jackal, and a troop of baboons. That morning we lingered, drinking two cups of Kenna coffee apiece. Archie drank sweet tea. The last few days had been brutally long and hot. Among the rocks Sonja and I kicked a ball, and Carol and Archie joined us. The sun grew warm. Soon Sonja retreated to shade. We followed. Archie paged through a *Time* magazine, two months old but new to us and Carol looked at a recent *Femina,* its pages filled with the latest in South African fashion and global gossip. Advertisements for American tobacco without health warnings, Tiffany's juices packed in ice, and green vegetables drew her thoughts elsewhere.

Gray bodies and coarse skin along the river offered a rare treat. Archie stared at elephants through the spotting scope. The equipment fascinated him. Skin magnified 60 times looked like a checkerboard, eyes glowed like fiery balls. Calves frolicked as adults crushed mopane leaves between molars the size of fists and stared toward us. We named this spot Elephant Camp.

This had been a marvelous trip in a very special area. The Kaokoveld, the northern Namib Desert, the southern Kunene Province, was a nearly waterless utopia, a desert paradise nurtured by dry rivers.

At bedtime that night Archie opened up, talking for nearly 90 minutes. He described his first meeting with Rudi Loutit, somewhere along the Ugab; he didn't know the year. He had wanted a job. Rudi suggested that he try his wife, Blythe. She'd wanted to know whether he had an approved driver's license, what the highest grade was that he'd completed, and if he could track.

Once employed by Save the Rhino Trust, Archie had delivered supplies to game guards in remote canyons. He drove an overloaded bakkie and attended meetings, including a meeting of the Wildlife Society of South Africa near Durban. Talks there centered on small animals, lizards, birds, and insects.

"It is good, it is good," he told me. "I can learn a lot."

Responsibilities mounted. Archie met dignitaries and guided them at the request of Save the Rhino Trust. He ferried them to parks, Etosha and Waterberg Plateau. He spent time with them in Windhoek and made airport runs. Patrons of the Trust were taken to see desert rhinos—people who had donated money to sponsor individual rhinos.

As Archie became better known, pressures increased. He saw affluence. He recognized poverty, his own. Two hundred fifty rand a month bought very little.

He didn't see his family often. Friends hit on him to use the Save the Rhino Trust vehicle. He discovered vodka.

"I drove drunk too many times to know," he mumbled toward the ground.

After one three-day binge he took a friend to an area to see rhinos.

"Then I saw the gun and didn't know what to do."

The rhino was shot three times. They left it and returned hours later with an ax and then drove back with horns in hand.

"I'll kill you, Archie, if you tell anyone," his friend said.

In the darkness, the flames reflected in his eyes. We sat quietly, neither of us talking. When Archie started talking again, voices from Sonja's tent muted his story. I struggled to capture his disjointed phrases.

After the poaching, he had devised a plan to kill himself while driving from Khorixas to Twyfelfontein. He didn't. Instead, he confessed. He was placed on leave, his position, his power, his dignity stripped. He had no salary. His father was disillusioned. A witch doctor from Botswana visited and administered emetics. They must have been effective, because Archie no longer drank.

He stood, exhausted from describing what troubled him deeply.

"Good night, Joel."

11

"The Missus"

[Carol]

almwag, a rustic guest camp with tall reeds and palm trees, sat at the edge of our northern desert site. Elephants regularly quenched their thirst at the bright oasis there before they disappeared among burnt red hills or into canyons. We too replenished our water supply, as well as petrol, and set up our tents behind the guest cottages. One evening, shouts and laughter echoed from the narrow entrance to the restaurant. As we started to go in, a stout, balding man blocked the passage. Staring intently at Archie, he muttered something unintelligible, and then moved his eyes to us.

In the entire room there was only one dark-skinned person, one black—Archie. Our friend took a step backwards. He didn't answer the man. His eyes flashed between Joel's face and the ground.

"Who are you?" said the man, switching to English to confront Joel.

"Visitors," was JB's curt reply; he met the man's hostile stare with an equally challenging look.

The man said something to me, undecipherable but still aggressive. *Maybe this is the owner,* I thought. Trying to be pleasant, I told him that we were rhino researchers.

"With whom—SRT?—I contribute a lot of money."

"No, we're independent researchers," replied Joel.

"Is he your driver?"

Joel said later that his initial inclination was to say, "Ask him. He can speak." Instead he offered, "He's a member of our team."

We, including Archie, slipped past him without further engagement and sat at a candlelit table to enjoy dinner. Soni played with two little plastic piggies she'd discovered among the natural history books and games resting on the corner shelves.

Back in Okaukuejo, the data chase continued as if we were characters in a bad detective novel. We didn't know what to believe. Blythe had told us that Malan was now in charge of the rhino database for all of Namibia. But when we asked Malan, he knew nothing about it. Then she'd told us that Hu Berry from Gobabeb, a research station nearly 1000 kilometers from Etosha, would have the missing records, but this proved to be another red herring. We telephoned and he knew nothing about it. Finally, we talked to Pete Morkel, the veterinarian on both de-horning operations. He didn't have the records of rhino identities or advice on how to find them.

We decided to try a different tack: We sought permission to enter the government vault and measure the horns themselves. Pete wasn't sure they'd be there. He said he'd heard something about a sting operation and the horns not being retrieved. Malan's account was similar, but he was sympathetic to our frustration and wanted to help. He and Eugene Joubert agreed to arrange for us to get into the Windhoek vaults. Perhaps we'd find the horns or copies of the original records, something to fill in the background and make our data meaningful.

Joel and I were asked by the ministry to make a formal presentation at their upcoming three-day Research Council Retreat (October 28–November 1) in Windhoek. Scientists from neighboring Botswana and as far away as England would also be there.

It didn't take long to work myself into an anxiety attack. "Joel, I can't believe I have to do this. You orchestrated the rhino project. I'm perfectly happy in the background. Plus you've given hundreds of talks. You know how nervous I get."

"Cunningham, if you don't speak, you'll fit expectations of the 'little woman'. You don't want to be just a wife, do you? Everyone should know you're doing more than just holding my hand and taking care of Sonja," he said. "Come on, Carol. Don't blame me. Just think about what you want to say and practice out loud. You can do it. Remember how back in South Dakota you lectured about the bison. You hated me for a week, but after you had practiced and it was all over, you felt good. You even liked me again."

Joel winked. It didn't matter that he was right. At that moment I detested him and the whole project. I didn't want to speak in front of audiences. Although I'd worked in biology for 15 years, I still felt uncomfortable, "on stage." I had a bachelor's degree, but in literature, not biology. Joel knew how nervous I was and regularly offered encouragement. He reminded me of my graduate courses in biology, even one in statistics, and told me over and over that I had watched rhinos interact at night, something that those in the audience had not done. His words were wasted. I was irritable. He had gotten me into this.

Other events proved powerfully distracting. I managed to relax between bouts of anxiety by watching animals. Although Damaraland had desert splendor enhanced by topography and color, Etosha's wildlife was more abundant, and they ignored people.

Ostrich chicks, bundles of grey feathers on spindly legs, appeared on our drives, racing after their single adult caretaker. Groups of 15 and 20 dashed from the roadside, high-stepping and zigzagging in a confusing jumble of legs and down. We had no idea if all of these babies came from a single pair of birds or if both the male and the female tended the nest or nestlings. Africa was so new to us. Brian Bertram reported in his book, *The Ostrich Communal Nesting System*, that several females actually lay eggs in one nest, but one pair, a male and a female, do most of the work. The major hen, the one who laid the first eggs and attended the nest, moves away from the nest and waits nearby when other females come to lay. What was astounding was that a nesting female could distinguish between her eggs and those of others. If the nest had more eggs than the pair could properly incubate (16–20), the extra eggs would be pushed out from the center and ignored, exposed to the blazing sun. Almost always, the sacrificial eggs belonged to minor hens, not to the incubating pair. It was remarkable that they could tell the difference.

There were also lizards, the most impressive in size being the white-throated savanna monitor. Big and carnivorous, the males weighed in at eight kilograms and had home ranges as large as 54 square kilometers, strikingly unusual for cold-blooded animals subsisting on snakes, lizards, and rodents. Dr. Andy Phillips of the Zoological Society of San Diego had captured 238 of the leguaans, placing radio transmitters on 31. Most astounding was that, collectively, the weight of the park's monitors exceeded the combined weight of all leopard, cheetah, and jackal living in Etosha at that time.

For a little more than a week we worked near Okaukuejo, maintaining nightly vigils for animals at the lit water hole. During the day there were chores. Unpacking, organizing, and washing our clothes, embedded with grime, by hand took several days. We'd taken to wearing old, tough T-shirts and shorts in the desert, changing them only when they looked beyond cleaning. Occasionally small items were washed while we were camping, but rarely, because the water was needed for drinking, cooking, and cleaning dishes. Not much was spared for faces, hands, and rinsing after brushing teeth.

I carted clothes to the ablution block and scrubbed everything, rubbing cloth against other parts of the same garment, vigorously dragging it across the ribs in the metal sink; if spots remained, full-strength soap powder and a small stiff brush usually finished off the job. Regularly, after camping for a month, I'd wash for two to three days in a row. Each day my arms and hands tired after a few hours and couldn't tolerate the strong soap, so I had to stop. Only the hardiest materials survived this beating, and rather than abuse more items, we used up as few of our clothes as possible. The nicer clothes that we carried for special dinners at rest camps were hung to dry outside the caravan, using equipment straps as clotheslines. Some items that we hung outside the washroom had disappeared. Occasionally, a woman named Adelina did our washing. Although she had another job, she always seemed to appreciate the extra money.

Adelina's help freed me for other chores in Okaukuejo. At the Ecological Institute I developed negatives of rhinos and made prints, and JB and I sat together for hours comparing photographs and confirming the identifications of individuals. I had taken photography in college knowing that it would be useful and because of a long fascination with photography as an art. We used high-speed black-and-white film that could be "pushed" when developed to function as if it were more sensitive to light, having an ISO rating of 3,200 instead of 400.

Our rhino pictures were slightly different from those taken by park management. When I printed, I made sure the enlarger was always at the same height so that all photos were standardized. Later, we measured the photographs, and because we knew the exact distance between Joel and the rhino, we could insert the measurements into a formula and estimate the actual size of the rhino or its horns without having to immobilize the animal.

The darkroom was a welcome hideaway, air-conditioned because warm temperatures damage photographic paper and chemicals. It was quiet there. Sonja was with Joel. The only sound was the water rinsing the finished prints. Working alone gave me a chance to relax, and considering the incredible amount of time we three spent together, a few hours apart had to be healthy. The management staff at the Institute had been wonderfully generous in letting me use the room, and we tried to repay them by donating supplies and photos and data.

Throughout this Okaukuejo stay, I never forgot that we were heading to Windhoek for the conference at the end of the month. We discussed who would present which segments. I'd give a little background and explain our choice of study sites and the mechanics of our study. Joel would round out the discussion and mention that he hoped to radio-collar rhinos to get a better idea of dispersal and home range size. As the date for the talk came closer, my nerves were calming down. I actually did like Joel again.

Coaching me, Joel insisted that we shouldn't rely on more than a few notes, key words at best; we'd present much better and actually be less nervous. He described how some speakers read their entire essay and were hardly heard or noticed by their increasingly comatose audience.

Our Okaukuejo routine solidified. Work, and play with Sonja, filled the day until about 5 P.M. Then we took turns running on the airstrip. I practiced the talk there, smoothing transitions and the order of topics, repeating them aloud as I scanned for lions. We returned to the caravan for dinner, made observations at the water hole, and retired late in the evening. I began to feel that I could do the talk.

There was still a problem to solve. We didn't know what to do with Sonja during the three-day conference. We asked at the Institute if anyone knew of a ministry person in Windhoek with small children. Day care or babysitting must be possibilities. Perhaps Soni could meet them in advance to make it easier on her. One downside of our family togetherness was that she'd only experienced day care briefly, and that was during the last few months in America. Here in Namibia, where everything was so different, we were anxious not to frighten her more than absolutely necessary.

Someone stuck his head into the lab where Joel was working. "Telephone call, Joel. You can take it up front." Dr. Rob Simmons, a British scientist working for the ministry in Windhoek, was on the other end. Apparently he and his wife, an American biologist named Phoebe Barnard, had recently seen a paper we'd published on bison and were surprised to see our address listed as Etosha. They wanted to meet us. Not only that, they had a daughter just 10 months older than Sonja. He suggested that we stay with them in Windhoek for the conference.

We followed directions and pulled into their driveway, angling wide with the Land Rover to avoid trimming the lower branches of a large acacia tree. Two dogs barked and bounced on the other side of the black iron gate. In a soft voice I said "Nice dog," and opened the gate as we drove in.

A little girl with pale yellow froth for hair came running across the grass yelling "Hello, hello!" An attractive woman with long thick hair, who had to be Phoebe, stepped out and welcomed us with a wide smile. This was as close to being home as we could have imagined. We ignored their accented voices and the gray lorries calling from the thorn trees. We felt at ease, happy to be there.

We traded histories; they told us stories of southern Africa, where they had been for eight years. Soni and Catherine laughed and played, declaring themselves "sisters."

Phoebe had arranged for Sonja to join Catherine's day care group. Anne Marie, the woman in charge, was German but spoke some English; leaning against the door frame with a baby on her hip, she assured us that Soni would be fine. We shouldn't worry.

When we finally reached the conference, we were exhilarated to be learning more about the country. Faces and names were overwhelming at first, but it was a good introduction to the ministry's varied personalities and perspectives on research and management.

Late that afternoon, we anxiously went to collect our little girl. When we arrived, she was sitting quietly outside, not playing and looking a little numb. When she saw us she burst into tears. Where was Catherine? She was there, but engaged in a game with some other children. We knew Catherine couldn't entertain Soni all day, and we felt guilty that we'd have to leave our sweet baby. Still, it was only for a few days, and we hoped Sonja wouldn't be too traumatized.

That night we again enjoyed the Barnard-Simmons household and discussions about ecology, conservation, and their research, which stretched from Canada to South Africa. Feeling last jitters before my talk the next day, I excused myself and paced the little patio outside the kitchen, practicing to no one in particular.

The presentations I'd witnessed earlier in the day had boosted my confidence. Most talks had been interesting and clear, but at least one person had stared at his overhead projector's sheets as if he had never seen them before. I knew I had to run through mine out loud a few more times.

I inhaled the fresh evening air. The two dogs eyed me curiously but soon lost interest when they saw my empty hands. *Sorry, no treats.* I stared blankly at the white brick wall and began rewording my introduction. But I knew that was dangerous and returned to the more familiar version. Finally I felt brave enough to rest. Unless I panicked, I felt sure I wouldn't embarrass myself.

At last the time came for our presentation. I told myself it would soon be over and I could just enjoy the novelty of the conference. Mannie Grobler, a biologist from the Caprivi Strip, began to introduce our talk. I couldn't concentrate. I breathed deeply.

Then I heard, "And now we will hear from Dr. Joel Berger." *We are both on the program. Why wasn't I mentioned?* As if in a trance, I stood and moved toward the podium. My heart beat wildly. A look of dismay came over Mannie's face. His eyes opened widely, and his jaw fell open. His voice faltered, but in a high-pitched squeak he managed to utter, "The missus." I had to laugh. Everyone did.

When I reached the front of the room, I introduced myself and felt good about the levity. It relaxed me at first, but I felt myself tightening up during the first few "official" sentences. For some reason, that was the hardest part. Once I began to tell about the different study sites and why we had chosen them, it went well. Joel followed and was entertaining. He also dropped one small bombshell—that after nearly seven months of trying, we still hadn't been able to locate the records of the dehorned rhinos. The tack was risky, but we hoped that mentioning it in front of ministry staff might facilitate matters. Eugene Joubert tried to rally support for us with his first question, asking if the data were simply misplaced or if there was a deliberate attempt to prevent us from seeing them. We didn't know, we said. At least now our major problem was out in the open. No one could say we'd never asked.

Two days later we had an appointment with a bank manager. Soon we'd be examining the government's collection of rhino horns hidden away in a locked vault. Guards escorted us down white halls into a darkened gallery where trunks full of horns were piled into a narrow room. Some were from animals dead from natural causes. Most were recovered from people who had shot animals and hacked the horns off for money.

We set about our task, measuring horns, recording sizes and weights, checking numbers and names, dates and locations. Some were from the 1989 dehorning operation, others from 1991. We checked and rechecked. To our amazement the horns from both dehorning operations were there—all of them. They had not vanished in a sting operation after all.

The day before, we had met with the Namibian Rhino Advisory Committee. On our behalf Deputy Minister Ulenga had asked Blythe for the missing data, pointing out that the information was to be shared with the ministry and us. Now, coupled with our finds from the vaults, we would have enough information to sort out the history of these animals.

Our trip to Windhoek had been a success. The conference had gone well. We'd visited with many of the ministry's staff and knew where they fit in the ministry's organization. The ministry had achieved its goal of encouraging dialogue between its research and management staff. And I felt I'd made a good showing as research partner. A personal highlight had been meeting Rob, Phoebe, and Catherine. We were sorry when the time came to climb back into the Land Rover and head north.

■

Year of the Tsongololo
[1992]

12

A Caprivi Crossing

[Carol]

The rains of the wet season drenched southern Africa. They began in November, moving first from the moisture-laden Indian Ocean and then toward barren lands further west. Soggy and inaccessible, the African veld closed its steamy doors to the outside world. Etosha rhinos, elephants, and other water-dependent species drank wherever rain pooled, disappearing into remote areas vehicles could not reach through thick mud and floods. Our most productive fieldwork took place in the dry season, when concentrations of animals or nighttime data were almost guaranteed. While the rains continued, we decided to visit Janet, Joel's Ph.D. student in Zimbabwe.

Our projects were coordinated. Her study of dehorned white rhinos in Hwange National Park had been underway for six months, and she wanted to show us her study area and animals. Janet had established a permanent base camp and thought discussions with Joel would be useful. We threw clothes into packs almost as quickly as Soni had unwrapped her Christmas packages, and we headed east through Etosha. Paved roads waited in Zimbabwe but not in Botswana or along Namibia's Caprivi Strip. Some 600 kilometers of mud, potholes, and rain provided an obstacle course.

Near the town of Tsumeb, enormous baobab trees with trunks appropriate for a giant-sized world appeared. Gray, vertical cords molded the main body of each tree and spread outward to form thick branches. Beyond Rundu, the main road paralleled the Kavango River, its edges hidden by a vast belt of tall grasses and reeds. Mixed riverine forests of fig and sausage trees disappeared into a colorful sea of teak and other woodlands.

The Caprivi was named in 1890 to honor a German general, Count Georg Leo von Caprivi di Caprara di Montecuccoli, Bismarck's successor. Little was known of the region when Germany assumed control, nor of the local people whose land was signed away. It was rumored that Germany had traded Zanzibar to Great Britain for the strip and could now access southwest Africa via the Zambezi River.

Although Joel had crossed the area previously, I was anxious to see the rich wildlife and water others had described. Vast, moist, and green, the region was flat and very different from the dry mountains and canyons we called home.

We explored lush waterways, detouring from one small haven to the next, some with recent signs of buffalo, others lagoons filled with turtles and frogs. The names of local rivers baffled us. Each country had its own name for the same river, reinforcing our difficulty understanding this exotic part of the world.

The "Cuando" River began in Angola but its spelling changed to "Kwando" at Namibia's northern border. Some 40 kilometers to the south the river reached Botswana's western bank and became the "Mashi". When the flow swung north back to Namibia's Mamili National Park and crept toward Lake Liambezi, it was the "Linyanti". And on the final leg to reach the powerful Zambezi, the river became the "Chobe".

We headed down a muddy dirt road to a small outpost known as Susuwe. Here government workers offered a presence that might discourage poachers. No one was at the rustic headquarters when we arrived, so we left a note and followed a narrow path around back to a porch facing the river. Our calls went unanswered. All was quiet except for an occasional bird and grunts that rumbled up from the Kwando River. We left and drove farther into the reserve.

Our wheels churned down a sandy track toward the grunts until we found hippos bathing in front of us. Real hippos. We could hardly believe this was still Namibia. Bulbous heads lifted and submerged—one large, another enormous, another half its size. They bobbed like floats newly released. Water flowed down rounded cheeks back into the river, and compact ears fluttered and rotated. It was impossible to judge the size of their submerged bodies, but we knew they were larger than black rhinos.

Sonja rode Joel's shoulders for a better view while I insisted they stay clear of the water's edge. Male hippos are notoriously aggressive, and have attacked boats and killed people. At night both males and females feed on land, where native and nonnative people alike give them even more respect. I nervously watched my fun-loving husband and daughter near the water. Although I loved watching animals in their natural setting, I was pleased when they returned. He reminded me that CC must stand for Cautious Carol, not Carol Cunningham.

We continued north toward the Angolan border. Near the river, trees and brush were thick. Hippo trails and droppings were everywhere. So were yellow-billed kites, egrets, and hooded cranes. We glimpsed several antelope haunches bounding through greenery. Waterbuck or lechwe? Our look was too brief to be sure.

Then our paradise ended. Tsetse flies were everywhere. Like warriors trained for hand-to-hand combat, fast and furious, they darted into the Land Rover. "Close the windows!" we screamed. "Quick, the vents, close the vents." Their bites were painful, and they were tough, staying alive despite our direct smacks. They had to be crushed against a hard surface. With strategic perfection, they hid silently and seemed to wait until we thought the increasingly stuffy vehicle was finally free of them before they renewed their attack. At last the flies inside the vehicle were dead, and we'd driven beyond the area.

Some interesting and much less obnoxious creatures were large dark millipedes. To our ear, the name seemed to be chongololas, or songololas, depending on who we talked to. Tsongololos are widespread in southern Africa except for in the truly arid deserts; some reach nearly 20 centimeters in length. Slow-moving, primarily consumers of detritus, to protect themselves they exude a cyanide-based repellent; in some species it burns the human hand.

We were drawn to a pretty spot at a bend in the Kwando River, one with deep winding bows and crystal water. The site was idyllic for an early dinner and rest. After our usual on-the-road throw-together pasta, I gathered the dishes for cleaning in our red tub. Joel entertained Soni, swinging her and dipping her legs closer and closer to the river's water. Instead of hippos, I kept thinking about crocodiles and scanned the water as I packed away utensils. Suddenly I was aware of a presence. A voice startled me.

"You're not planning on staying long, are you?"

My head jerked around. A uniformed man with an automatic weapon draped over his shoulder stood five feet away. Two other armed men stood at a nearby tree, quiet and expectant. Joel walked up slowly from the river. I explained who we were and that we had stopped at headquarters. In crisp, clear English Simon Mayes, the leader, revealed their predicament. They had caught a poacher. After questioning, they had learned that more poachers planned to cross the river from Angola. An ambush had been set for tonight. Its location was the very bend in the river where we dined. Politely, he suggested that we move.

"Of course, of course." We smiled. Then, with ungainly speed, we tossed our things into the vehicle and raced back down the spoor. The soldiers had already disappeared to await their prey.

We pulled onto a sandy track and looked for a new spot to camp. Around dusk, hippo snorts and grunts came from the river below. We slept soundly on the Rover's roof until, at 5:30 A.M., thunder and rain woke us. We pulled canvas over ourselves, and the soft patter of falling drops lulled us back to sleep.

The sound of gunshots interrupted breakfast. *The ambush.* Unsure of the situation, traveling unarmed, we gathered our things and drove toward the main road, hoping no one would stumble on us.

Three barefoot men blocked the muddied track. One wore a uniform of sorts, black pants and an unbuttoned white shirt partially tucked into a waistband. We couldn't tell if they were poachers or rangers. They lacked the usual green government attire, and they refused to move. There was no way to drive around them; Joel hit the brakes, and the Land Rover ground to a halt.

Odd syllables tumbled from their tongues, and arms gestured wildly, until it was finally obvious that we couldn't understand. In broken English one man told us that they needed a ride. I glanced at Joel. He didn't know what to do either. Could they be poachers escaping the ambush?

"What if they really need help, and there is no connection to the gunshots? Should we give them a lift?"

Hardly any villagers had cars. These men seemed nonthreatening, if in a hurry. With Soni up front, we nodded yes and pointed to the crowded back seat. Legs and arms folded against trunks as the three squeezed in. Water jugs, fuel canisters, and duffle bags had already filled any spare legroom. We tried to talk, while we drove, exchanging halting conversations with few words. They guessed we were American.

Finally we understood that they were trying to help some people whose car had been crushed by an elephant. It was a wild but believable tale considering the region and other stories we'd heard about rampaging elephants.

Suddenly a government truck careened over the top of a hill. Joel and the other driver swerved to opposite sides of the track and stopped. Our riders jumped out, followed by the uniformed occupants of the other vehicle. Arms waved, accompanied by unintelligible shouting. Waiting and wondering, we hung back near our car. Abruptly the exchange stopped, and the others leapt into their vehicle and drove off. Our guys remained. We were stunned.

Again they squashed into the rear seat. Unsure of what had happened, or what was supposed to happen, we just kept driving. The main Caprivi road had to be ahead. When we finally reached it one man tapped Joel on the shoulder, and all got out of the car, smiling and waving. It was over.

We returned to the Susuwe headquarters to report the shots, the men, and the possibility of a crushed vehicle. This time a young man in the familiar khaki shirt and shorts walked toward us. Immediately we began to babble about gunshots and barefoot men and vehicles. Then we took a deep breath, stopped rambling, and started over.

As we talked to another government worker Matthew Rice, over tea, a more coherent story emerged. He'd heard about us from Simon the previous night and had been concerned about an international incident if Americans were shot in a skirmish along the Angolan border. As it turned out, no poachers had even tried to cross the river. Simon was already out investigating the gunshots we'd heard.

We left, traveling east toward Mdumu National Park. At a police check station, several officers approached. One leaned down to the window and greeted us enthusiastically, "Hello, hello." It was one of our early morning hitchhikers. We passed without interrogation.

* * *

When James Chapman traveled the Caprivi in 1868, he found an area "where the elephant is not much hunted, or firearms are almost unknown, as at Linyanti, they come in herds to the cornfields in daytime and commit great depredations, to the terror and grief of the natives, who are often starved in consequence." By 1991, some 50,000 people filled the villages of the eastern Caprivi. Among them were refugees from Angola, visitors from Botswana, and a few tourists hoping to see wildlife. People lined the muddy road. Women balanced pails overflowing with possessions, baskets, and even suitcases on their heads.

It had been only 15 years since Caprivi was a mecca teeming with wildlife, and it still serves as a migratory passageway to Angola, Botswana, and Zambia. Nowhere else in Namibia are there bushpigs and swamp-adapted antelopes like puku, sitatunga, lechwe, and waterbuck. But during the war between SWAPO and South Africa, wildlife along the rich watery corridor suffered. Buffalo and elephants were slaughtered indiscriminately. In 10 years, the lechwe population dropped from an estimated 13,000 to less than 1,300. By 1991 only five puku had been seen. At Mamili National Park, the tsessebe, plains zebra, wild dog, waterbuck, and black rhino were extinct.

* * *

At a large elegant table, we dined with German, South African, and British guests at Lianshulu Lodge, set within Mdumu National Park. An exquisitely prepared meal, pleasant conversation, Soni eating and behaving well—I was enjoying our stop wholeheartedly. The tranquility was marred by the arrival of a large narrow beetle as long as my thumb. It swooped down toward the candlelit table and landed on my lap. I had become so accustomed to the relatively gentle dung beetles zooming in and out of our caravan at Okaukuejo that I didn't even look when I picked up the beetle. I shrieked an undignified "Ow," with more volume than if I'd been thinking about manners and the gentile setting. Blood poured from the spot where I'd been stabbed.

Sonja enjoyed a grand dinner, far better than anything she could expect from her parents. Afterwards, we all lay cozily on a huge bed; mosquito netting hung from the high, thatched ceiling.

The next morning we explored the Kwando waterway in a small boat—

This was the tip of the Linyanti Swamp. From the low flat-bottoms we saw birds: dramatically colored flycatchers, white-fronted bee-eaters, rufous-bellied herons, pygmy geese. Dainty lesser jacanas moved fleetly across lily-covered ponds. We even glimpsed a rare western-banded snake eagle as we floated past a tall tree. Sonja giggled as Joel dipped her feet up and down into the winding river until a crocodile slithered from the bank and disappeared into dark water; no more dipping for Sonja. Hippos raised their heads but submerged them quickly. Behind us, the enormous head of a male stayed up and eyed our retreat. He'd been highly successful at his job. No one was going to bother his family.

Although the Caprivi was a verdant paradise, for desert rats like us this rich waterfed ecosystem had drawbacks. Besides tsetse flies there were malaria-carrying mosquitoes. One night on the Land Rover, we fastened the mosquito netting carelessly. I slept terribly and woke with some 400 bites. JB was spared: He had only 50. Within a couple of days my skin resembled sandpaper, corrugated and lumpy. It was impossible to tell where one bite ended and the next began. When we refueled, the three men at the petrol station couldn't stop gaping at my disfigured flesh. We hoped our antimalaria tablets worked.

Tall green forests greeted us as we entered Zimbabwe—formerly Rhodesia under British colonial rule. Its eight million citizens were free now, thanks to a bloody bush war that ended in 1980. Zimbabwe's commitment to conservation is exemplified by an economy that is partly tourist-based and dependent on wildlife. Besides Victoria Falls, parks like Mana Pools and Matusadona are scattered along the Zambezi River Valley. For the adventure-minded tourists, there are undeveloped reserves such as Chizarira and Gonarezhou in remote areas. There are also accessible spots like Matopos near the city of Bulawayo.

Because of the wet environment, Zimbabwe's wildlife is more diverse than that we'd grown accustomed to in Namibia. Among the many species we hoped to see was the Samango, a relative of the vervet monkey, whose name is derived from the Zulu word iNsimango. These handsome primates, white-throated with black heads and shoulders, are found in groups of up to 30 animals. Samangoes live in Zimbabwe's lush eastern highlands, unfortunately a day's drive from Hwange National Park, where we hoped to find Janet Rachlow.

Janet's study of white rhinos was well underway. We hoped that she could show us her study animals. Compared to black rhinos, "whites" are massive. If not for Asian and African elephants, white rhinos would be the largest land mammal on the planet. Males can weigh in excess of 3,500 kilograms, their necks are thick and muscular, and horns have reached nearly 200 centimeters in length. (White rhinos are not white, but gray like black rhinos. The name is a misnomer, perhaps derived from the Afrikaans word *wyt*, meaning wide, in reference to this species's broad, square-shaped lips.) Unlike their "black" relatives, white rhinos are grazers—grass-eaters—and feed in larger groups, and more frequently in open areas.

The species was never very common in Zimbabwe or Namibia. First reported to the western world in 1817 by William Burchell, they were in danger of extinction in southern Africa before 1900. In Hwange, Zimbabwe's crown jewel of parks, the last native white rhinos were seen by Frederick Selous in 1873. A reintroduced population now occurs there and in a few other Zimbabwe parks.

We drove with Janet to her tent camp at the Linkwasha Vlei, a stunning grassy expanse surrounded by forested hills. She had already seen leopards and lions nearby. Sonja wouldn't be walking far without supervision. Toward evening hundreds of the park's 30,000 elephants congregated with buffalo at a water hole. Others fed nearby. As the light muted, two white rhinos appeared, feeding peacefully side by side with the buffalo. Then, without drinking, the duo faded back into the thick woodlands. Our brief glimpse revealed that they were hornless.

The Zimbabwean government had been busy dehorning their rhinos, white and black. Like us, Janet was studying the biological effects of horn removal, and she suggested that it was better for the government to harvest the horns than for poachers to kill the animals. We agreed. The government is serious about rhino protection; their controversial 1987 shoot-to-kill policy targeting poachers still continues.

As the days passed, we toured more of the park, seeing hippos, crocodiles, and sable antelope. Saddle-billed storks, Africa's largest, pumped their powerful wings and floated overhead while pygmy kingfishers caught ground-dwelling insects. Common impalas were everywhere. Baboons amused the tourists. We saw the tracks of wild dogs, canids that hunt in packs—Josh Ginsberg from the Zoological Society of London was studying them. Sightings reported by tourists helped his team locate these wide-roaming carnivores.

But there were other reports, alarming ones. Gunshots had been heard during the day from areas just outside Main Camp, the primary destination of many Hwange tourists. Poachers here were brazen. Four dehorned white rhinos that regularly grazed near the camp hadn't been seen for days, and now there was concern that even they had been poached. Searching with Janet in her Land Rover, we found only two of the foursome. The mother of a juvenile was missing. Another rhino now had a conspicuous limp; a bullet might be lodged within it.

If a dehorned animal had been shot, no one understood why. We speculated that, in thick vegetation, it was hard to see that an animal was dehorned. We wondered if the sliver of remaining horn was worth the risk of being killed. Perhaps poachers just didn't want to waste valuable time tracking rhinos without horns and found it more efficient to slaughter them. Perhaps they killed rhinos just to spite the government, to declare that these conservation efforts wouldn't work. Or perhaps continued poaching was an attempt to drive up the price of illegally stockpiled horn once the species was exterminated.

Fortunately, Janet still had plenty of dehorned white rhinos for her study.

The wet season would last a few months more, but we grew antsy. With hugs, wishing Janet the best, we headed back to Namibia. Reviewing the dreadful situation, we hoped that cutting the horns from rhinos would allow future Zimbabweans a chance to see their rhinos.

13

Rhino Illusions

[Joel]

Back in Namibia, bursts of rain punctuated summer's humid days. Mosquitoes still plagued everyone; while tourists used prophylactic drugs, most Namibians didn't, and many became sick and lethargic. The Etosha Pan no longer shimmered. Its surface was bone dry and muddy by turns. Flamingoes, which breed successfully once every four years on average, renewed mating attempts. The veld water was widely dispersed, remaining in a few humanmade dams and natural depressions. Waist-high grasses concealed rutted, two-track roads, some with holes half a meter deep, sites where elephants walked and sank. To reach our second study site within Etosha, an area we called Calcrete Basin, would be nearly impossible even though it was only 100 kilometers from Okaukuejo. Our time would best be spent on other activities.

The caravan that the San Diego Zoo had given to the ministry had become our infrequently used home. Tucked away at the southern end of Okaukuejo, it was shaded by green netting. Not far away, at the Ecological Institute, grey lorries, splendidly bright wood hoopoes, and chameleons found refuge on immaculate lawns. Cobras enjoyed the plants and coolness. After being startled by moving shadows, we regularly used a torch at night.

As in most arid places settled by humans, their demands for water have shaped the landscape. To combat Okaukuejo's dropping water table, conservation measures were implemented the year we arrived. However, we were surprised to see water flooding the roadways as sprinklers soaked lawns and gardens. The hose at the petrol station ran perpetually, and tourists, ignoring requests to limit water usage, borrowed the hose to wash dust from their cars. Yet in the bathrooms water was turned off for hours at midday. Toilets couldn't be flushed, and excrement literally piled higher and higher. Open urinals smelled, and flies and mosquitoes buzzed contentedly.

During the day we crunched data, and Carol also worked in the Institute dark-room. Sonja, now two and a half, alternated playtime between us but she also enjoyed her new friends, Pieter and William, who lived at the opposite side of Okaukuejo with their parents, Wilfred and Elsie Versfeld. Although initially the boys spoke mostly Afrikaans and little English, the trio had fun, kicking balls, climbing trees, and chasing pet chickens. During hot afternoons they cooled them-selves in a wading pool. Language barriers mattered little. Elsie loved Sonja, the daughter that she never had, and made dresses, cookies, and conversation. We all became friends. Wilfred offered insights about life under South African rule. I did the same about life in America.

We also dealt with correspondence and paperwork. Progress reports to the ministry and our grantors were due. And just because I was in Africa didn't mean that I could escape my faculty responsibilities at the University of Nevada at Reno. Reports on committee actions, votes about bylaws or tenure, and input on course-work waited. Interesting requests about filming rhinos came from crews in Italy, Germany, and the States. We suggested they contact the ministry. Others were from potential volunteers, students wanting to pursue conservation careers, and family members and friends hoping to visit. The overwhelming logistical challenges of working in Damaraland—the need for another Land Rover because ours couldn't fit another body, and the requisite permits—meant that we wrote many polite answers: "Oh—it would be great to have you visit, but. . . ."

One cable drew our interest, a short message from a Dr. Franz Camenzind, a biologist turned filmmaker. He had photographed pandas, wolves, and grizzly bears, and had even been to Namibia. Although we hesitated to invite someone that we didn't know, we saved his request.

Wind blew in our faces, and dark clouds filled the eastern sky, as we prepared for a jog at Okaukuejo's airstrip. Lightning moved closer. A long-legged secretary bird with a red face bounced as it ran in short steps, flapping its wings wildly. Then the rain began, chipping slowly at the dried earth. Small rivulets pooled. The air was fresh. We alternated our runs with Sonja-sitting. Once the clouds parted we ate dinner on the runway, watching a fiery sunset. Our lives were rich. The three of us were together. Sonja was healthy. Carol and I were studying rhinos. We shared a special feeling. We were lucky.

The next morning, clear skies and a small airplane greeted us at the airstrip. Bill Gasaway also met us there and helped duct-tape antennas to plane's struts. Thanks to Pete Morkel and his untiring efforts during the past few months, three rhinos now wore radio collars. Our goal was to find radio signals from the collared

animals somewhere in central Etosha's thornbush. We'd plot locations and estimate daily, weekly, or monthly movements. Finding or darting the animals was
not an especially serious problem. It was exciting.

Once, while Sonja enjoyed a midday nap in the Land Rover's back seat, a
recovering rhino with a radio-collar stood and charged. I dashed behind the vehicle. From the front seat Carol sat wide-eyed as one thousand kilograms attached
to a half-meter-long horn shot toward the back seat where Sonja's little head
pressed against the door. Metal collapsed; the horn connected just an arm's length
below Sonja's head. Then the rhino disappeared into the veld.

Our major dilemma was keeping collars on the animals. Rhinos have massive
necks, thicker than their heads. To escape the collar, all a rhino has to do is rub
or hook it on a branch. With the ears all that kept the collar in place, neither
Pete nor I were anxious to have collars fitted too tightly or to use hard material
that might damage the sensitive ears. At Pete's suggestion we used pliable seatbelt
webbing to avoid abrasions.

In South Africa radio implants had been placed in chambers drilled into the
horns, a tactic we decided against. Home ranges in the thick moist forests there
were much smaller, generally less than four square kilometers, and the horn implants released signals that could be heard within that range. With neck collars,
our detectable range on the ground was only about three kilometers, but from the
air we could hear signals from 15 kilometers.

Our first radio collar lasted only about 10 weeks before the female shed it like
a snake shedding its skin. Our next victim, an old male named *Apache,* was darted
near Zebra Pan. He was a mild-mannered bull, subordinate to males and petrified
of females. His collar remained on for nearly 12 weeks before slipping off in the
Dolomite Hills to the west.

Since poaching had been down in Etosha, I convinced the management staff
to let the Anti-Poaching Unit (APU) work with us for a week. The plan was that
Carol, Sonja, Pete Morkel, seven APU members, and I would all camp and track
rhinos together. The APU guys—Matheus, Mauo, Meuus, Shipero, Jonas, Paulus,
and Daniel—spoke Owambo, Herero, or Afrikaans. No English. Their leader, a
war veteran from Zimbabwe, spoke several languages. We ironed out the details.

I'd cover expenses and food. Alcohol was prohibited even at night. Carol and
I would have to find a different way to express our appreciation. We decided to
pick up frozen steak fillets from Okaukuejo's store, something that should be a
real treat given how rarely these men ate red meat. We delivered them to the APU
leader's house and agreed to meet at the horse camp in the west around sunset.
The meat would have plenty of time to thaw for tonight's dinner.

These large operations were intensely stressful. As during the dehorning effort,
there were always too many personalities and too many chances for things to go
wrong, not to mention the safety of the rhinos and equipment-related problems.
On the drive, Pete Morkel indicated that he had brought two tins of beef for the
trip.

"Two, for five days," whispered Carol, shocked.

It didn't matter. We had plenty of food, and the team leader was to bring the supplies that we had just bought.

When we arrived, cast-iron pots were draped across a smoldering fire, and the smells of burning mopane wafted in our direction. The men were already eating, but there were no steaks.

"The men," said the leader, "all thank you for the Cokes, and they are anxious to find rhinos for you."

Cokes, I thought to myself. *What about the steaks?*

"Didn't you give them the steaks? I want them to know that this is a special occasion. I appreciate their willingness to help." What I wanted to ask was what the hell was going on. I knew better and bit my lip.

"These men shouldn't be eating steaks and getting used to having things given to them, especially by foreigners. They'll get soft." The leader turned and walked away.

During the night it rained. Finding rhinos at this time of the year was difficult because many moved away from the over-utilized vegetation near water holes to areas with both veld water and plants with new growth. Nevertheless, collaring more animals was our only hope of getting better estimates of home range sizes and seasonal changes.

Early that morning the APU team picked up the muddy spoor of a solitary adult. Carol, Sonja, and I followed, bashing the Land Rover through brush up to four meters high, scraping and dragging branches as thick as fists along the undercarriage, and plowing through deep mud. With each new obstacle, we wondered which wires were being yanked free.

Three hours passed without a rhino. Confusion reigned. The 21-centimeter footprints that we tracked merged with others of the same size, then separated, and continued walking in a virtual straight line. This rhino had a mission. It was not feeding, just walking.

Another hour passed. The animal had covered 19 kilometers from where its track was first found, but its large footprints were now accompanied by small ones. The animal we followed must be a female, now united with her calf. Twenty minutes later we saw them, a cow and calf fleeing from our army. We called off the forces.

When this project ended, we headed back to our Zebra Pan site.

"Graviportal . . . vertebral column balanced on the forelegs, with the head counterbalancing the body weight and the hind limbs providing the main propelling force, the wide-set, pillar-like limbs and 3-toed feet making a firm foundation." Thus Richard Estes described Africa's two rhinos. He also suggested that these animals are not as asocial as had been typically supposed in the scientific literature. He was correct.

Early travelers found rhinos to be semisocial, congregating in groups at water. Watering sites were found littered with rhino bones and skulls. British naturalist Francis Galton reported on his 1851 travels which included Namibia and Botswana: "Rhino skulls were lying in every direction. . . . Saul had told us that the rhinoceros would begin trooping in at nightfall, and that we should continue firing at them till daybreak, and I believed him. Forty were killed here about a month since. I could not doubt it, for I counted in a small place upwards of twenty heads."

Near Ghanzi in Botswana, Galton's traveling mate, Charles John Anderson, wrote: "One night, I counted twenty defiling past me, though beyond reach."

In 1860, explorer Richard Burton indicated: "black rhinoceros are as common as elephant" in East Africa.

The favored way to kill rhinos in southern Africa then was to wait at night for the thirsty animals to approach the water holes. The tactic had obviously worked well, and we now adopted it. Only our aims differed. We came for data, not blood.

Our observation periods at Etosha had three goals. First, we needed to identify every animal each night. When individuals had distinguishable marks and had previously been photographed, we merely watched without disturbing them. But there was no doubt that our presence had an effect. Some would hesitate to drink. With deliberate pause, they sometimes approached our car and circled, often in jerky movements with explosive outbursts. Gradually, the animals habituated to us. And whenever another rhino approached, they ignored us. If the individual was unknown, or we weren't certain of its identity, we photographed it at close range. Not only did we now know who it was, but we could later calculate its horn size to determine if it affected social status or defense. When possible we'd gather urine or feces left by females for later pregnancy tests. So, although our first goal was identification, the same data would also be useful for population monitoring.

Our second and third goals concerned rhino social biology. We wanted to learn about the relationships between rhinos themselves as well as how rhinos interacted with potentially dangerous animals. The latter included lions, spotted hyenas, even elephants. Therefore, we simply sat, gathering data that could often be very confusing. One morning at about 2:30 A.M., five individuals came to drink virtually in a single file. None of them had been among the 11 who had visited earlier in the evening. When the quintet arrived, two others were standing in the center of the opening, half a kilometer from the thick woodlands. Squealing, roaring, and snorting ruptured the silent darkness. Heavy dust, the result of deliberate foot shuffles by rhinos, obscured our view. We concentrated on the two animals with the most distinctive horns and ears to avoid confusing individuals.

Sudan, Zambia, Kenya, Angola, Botswana, Zaire, and other African nations were the names we chose for the rhinos in the Calcrete Basin population. Subtle physical variations distinguished individuals. *Angola's* left ear was missing its tip, and he had a horizontal tear across his right. *C. A. R.,* for *Central African Republic,* had a broken posterior horn; *Ethiopia's* posterior horn was cracked at the top, *Gabon* was missing part of her tail, *Kenya* had lost portions of hair tufts at the top of her left ear, *Mozambique*'s ear had been notched during a transplant operation, and *Zimbabwe* had three horns, two normal ones plus a small knob centered above his eyes. We knew more than 25 animals by such signs. Others were more difficult to identify, even with the aid of our photographs, and were not easily distinguishable from one another.

We were not sure that Calcrete Basin would serve our data needs. Driving there for the first time was like entering a different world. Unlike the openness of the Namib Desert or the short thornscrub of the Zebra Pan area, the basin was within forests of red bush-willow, leadwood, and purple-pod terminalia. Most trees were five meters high. The mopane trees were even taller.

We had been told to look for a large open area nearly 300 kilometers from Zebra Pan, but as we lurched around elephant holes and rocks, we were skeptical that a vlei existed anywhere within this dark forest. Then suddenly there were no trees, and a tall wooden column stood on a naked plain. It was an observation tower. Trees were nearly half a kilometer away or farther. The area was perfect. A broad pool formed by water rushing from an artesian spring was surrounded by a hardened calcrete surface. With a full moon, Calcrete Basin was the arid equivalent of an arctic landscape. Just as snow reflects a bright moon, light bouncing off the white calcrete left us with the eery ability to see in darkness even without the image-intensifying night equipment. The wooden tower contained waist-high walls and a thatched roof. Panels extended from the roof to the walls that could be lifted to allow a breeze or lowered to block the wind. The tower would offer shade during the day, a place to store gear, and a refuge from lions and elephants.

A routine developed. At first light, hundreds of squawking guinea fowl left the protected forest and marched to the water. Wildebeest and zebras stared as a human body wriggled from a sleeping bag and descended the Land Rover's ladder. Making the coffee was my duty. Then I'd scan the vlei and woodlands for other signs of life.

My immediate goal was to check the spoor of the animals who had come to drink after we had gone to bed. Among the guests that we regularly missed were lions. More than once Soni's tiny footprints from the previous night disappeared under crescent-shaped depressions that led right up to the Land Rover. Only with nearby roars did we wake, lying stiffly and pretending to be part of the metal grating.

Sometimes we learned that additional rhinos had visited. Diminutive footprints might accompany those of a mother. Since we hadn't seen calves of that size during our nightly observations, we surmised that they must have appeared after our four to five cups of coffee failed to keep us awake.

During the day the tower was a welcome respite from the intense sun. We charged batteries with solar panels, organized data, and rechecked equipment. When we couldn't nap, we enjoyed the antics of the animals below.

More than 600 zebras might pass in a single day, squealing, chasing, or fighting. Males battled often, rearing up on their hind legs or biting savagely at each other's faces or necks. When the fighting turned serious, they dropped to their knees and bit viciously at their rival's lower tendons. Both combatants remained on the ground, legs tucked below their bodies for protection, looking like turtles too afraid to move.

The males of many ungulate species possess secondary sexual characteristics useful in competing for mates. Male elephants, kudus, and giraffes have tusks, horns, or thick necks that help to gain advantage during fighting. In zebras, however, the sexes are virtually the same size, and conspicuous male weapons seem to be lacking. Like male horses, male zebras nevertheless have piercing canine teeth in both their upper and lower jaws, specialized weapons not found in females. The canines may be responsible for leg damage inflicted during serious fights, injuries that could lead to death from hungry lions and hyenas.

We became familiar with our thirsty visitors. Gemsbok, wildebeest, and zebras were regulars. Sonja named one wildebeest Wilbur, a territorial bull who stationed himself nearby and became her favorite as he waited for females and plotted attacks on males. Red hartebeest and white, black, and tan springbok added color to the parade of diverse herbivores. Giraffes and warthogs were plentiful.

Among the rarer guests were large but curiously shy eland, and endangered black-faced impalas. Male ostriches waved long black feathers in gaudy displays to gray females. The cheetah visits were riveting. Zebras bunched and stared. Gemsbok lowered their rapierlike horns and followed, occasionally rushing the cats in short charges. Elephants shared in the diurnal extravaganzas, regularly displacing other species from the water, chasing zebras or tossing their trunks toward thirsty gemsbok. During one fracas, a warthog avoided a well-aimed trunk by diving into the water.

During the afternoon, bateleurs and other eagles sailed in on thermals. White-backed and lappet-faced vultures panicked the zebras into dusty escape attempts.

We passed the daytime lull with binoculars, books, and pillows as we waited for the night. Sonja practiced identifying all the mammals and could tell the difference between eagles, vultures, storks, ducks, guinea fowl, and plovers.

Despite the heat and our rising irritation from lack of sleep, we made special efforts to entertain Sonja. Besides watching animals from the tower, she swayed in a swing that I fashioned from nylon straps. Down below we'd gather wood, bake bread, and build dams at the water's edge. One day while we were out walking, she was the first to see a partial skull, eye sockets in place, and large recurved canines. "Daaa-aad, warthog," she exclaimed.

Although safe in the tower, we always tried to remember how well lions conceal themselves. When days or nights passed without roars or spoor, a false sense of security developed. To make matters worse, we told each other (mostly I told Carol) that the behavior of the ungulates would warn us if lions were near. This reasoning was obviously circular. Lions make their living not only by scavenging but by being successful predators. Prey are not eternally vigilant, and they do not always avoid attacks. Using ungulate behavior as a gauge was suspect. We really never knew how many times we failed to see lions either during the day or at night.

Our routine also included checking bones and rocks that we had neatly piled at the water's edge, markers to indicate distances from which I could photograph rhinos. At night everything looked different, and I memorized which rock cairns were associated with which bone piles.

Once the sun dropped, equipment was checked a final time. Two spotlights, both permanently fixed on the Land Rover's roof rack, aimed forward. Three cameras with different film speeds and telephoto lenses of varying size nestled against the rack's metal railing. Thermoses, crackers, and other snacks were to the right of the spare tires. Equipment cases served as tables, while two portable spotlights rested within arm's reach. On the dashboard were our Bosch adaptors from Germany that converted energy from our third 12 volt battery to 110 direct current.

We had too much gear and too many wires. To avoid entanglement we wound the cords around the tires and down the windshield. Extra sweaters, sweat pants, and other clothing for winter nights poked through the metal grating or hung from the side of the car. We sat on a mattress. Data books, cups, two backpacks, and spare batteries for our flashlights and cameras were included in the necessary pile. The roof rack was at carrying capacity. There wasn't much room to move once we settled in for the long hours of evening "obs."

As night fell, we knew Sonja was ready for her back-seat bed when she became restless and noisy. But after we emphasized the importance of quiet when animals were nearby, somehow she understood and settled down. While Carol read bedtime stories inside the car, I remained on top, motionless except for my scanning with the binoculars. Dim shapes and clumps at the forest's edge became jackals, cheetahs, and wildebeest. Sounds, too, revealed identities. "Sssslippp, sssslippp" indicated lions at the water. A pulsing wheeze was a springbok. Zebras brayed. Elephants were often dangerously quiet.

Rhinos appeared at the forest's edge like airplanes circling a field, waiting patiently for others to be scheduled before coming in for a landing. Photographing unfamiliar rhinos was tricky. They were very wary when arriving alone. And, without Carol's eyes for backup, I was forced to scan for lions while still concentrating on the rhino. I always felt vulnerable. Close range amplified the danger. Sometimes I lay on the dirt for up to 10 minutes less than 30 meters from a rhino just waiting for the perfect angle before shooting the picture. Being so close meant that if I tried to refocus my eyes to scan for lions, I'd never pick up the rhino in time if it moved toward me. My nerves were frayed. I always felt safer when Carol climbed back on top of the Land Rover. Four eyes were better than two. And, with Carol, a slight whisper or a powerful beam from a spotlight could alert me to approaching danger.

Cunningham's good eyes did not guarantee safety. One night, while I was out with three rhinos, another appeared. We both first saw it 75 meters behind me. There was no escape.

With rhinos moving here and there, it was impossible to keep them all in sight at the same time and still gather data on a focal animal. Carol whistled me back to the car and pointed out two nearby hyenas. Usually I didn't worry about hyenas if rhinos were around because the pachyderms often chased them away. But with the addition of a third and fourth hyena, I was less confident. One female rhino that we called *Female A* was also there. She had been skittish for months, and we still needed an identity and horn shots. Despite the chaos of four rhinos and four hyenas, I wanted her photograph.

With *Female A* near the other rhinos, I had my chance. She was distracted by them and less concerned with me. Binoculars drooped from my neck, and a camera and flash unit were in one hand. With a flashlight jammed into my belt and a battery pack dangling under my free arm, I inched down the ladder. A faraway roar jarred our nerves but also told us lions were still distant. Or at least some were. Carol continued scanning. The improved night vision equipment had a larger lens that magnified images up to 13 times. So, despite the darkness, we could see all the way to the forest's edge—600 meters in places.

Slowly I approached a rock cairn that was still 20 meters off. A large object moved in the shadows behind the rhinos—at least, I thought so—but as I peered into the night I couldn't see a thing. Anxiously I glanced to Carol for help, but she was busily watching something else, probably the hyenas.

Again, something moved. This time, terror struck. I froze where I stood. Less than 30 meters away a massive lion had materialized. He increased his pace and suddenly sprang into motion.

I wanted to run but couldn't. A lump developed in my throat. A millisecond passed, not enough time to really think about gruesome possibilities. Then, ignoring my upright body, he exploded toward the hyenas. I returned to the Land Rover. Carol's face popped over the side as she looked down at me.

"Did you see that lion? Good obs, Cunningham!" I rasped.

"I know, I'm sorry. I was concentrating on the hyenas and didn't see him until he went after the hyenas," she said.

We made progress toward two of our three goals. We came to know rhinos individually, and data on their social biology streamed in. Some nights there would be no rhinos for hours; then, uncannily, four or five different individuals would arrive at the same time. One night at least 17 different animals came to drink. Several times as many as 12 drank simultaneously.

It was never totally clear whether any single individual consistently dominated others. *Somalia,* a male with thick, worn horns and floppy tattered ears, visited every two to four days, usually in the early evening. He never ran from other rhinos or avoided interacting, and he rarely fought. Once, however, at 3:10 A.M., he chased a younger male, *Mali,* into the water and thrust his horn into *Mali's* side. *Mali* broke from the chest-high water, fleeing in a panic. *Somalia* caught him anyway, and they faced off. Squeals and roars reverberated. We never saw *Mali* again. Either he died from the wounds or found another place to drink.

Although males occasionally engaged in intense combat, females were consistently more aggressive. We never knew why. Although water was the common focal point for interactions, the pool at Calcrete Basin was large, and all the rhinos had access to water. Perhaps fighting was related to reproductive status.

Some females with large pendulous teats entered the woodland clearing and progressed to the water. No calves accompanied them; without the night vision equipment, we might have thought these animals were males, mistaking their teats for dangling penises. These females must have left their calves concealed in the veld, because six months later they appeared with young in tow, calves at least six months old. Maybe lactating females with young calves were more aggressive so they could finish drinking sooner and return to their hidden, unprotected babies. Only once did a female with a swollen udder not arrive subsequently with a calf. We assumed that her calf must have died because her udder had lost its pendulous appearance. Why was a different matter. Abandonment, predation, disease, or some other factor were all possibilities that we could not assess.

Nights at the Etosha waterholes were seldom dull. Few places remain in the world where so many rare or endangered mammals can be seen at one sitting. On good nights there would be elephants, rhinos, black-faced impalas, cheetahs, and brown hyenas.

By the time we went to sleep we were often emotionally and physically drained. We never knew whether elephants or lions would cause more trouble. Both species were unpredictable. When we saw elephants enter the clearings we usually climbed into the car and waited for them to leave, even if that meant forfeiting rhino data. Occasionally, we'd even start the engine, slide the car into reverse, and back away. Knowing that Africans and tourists die from elephant attacks, we accepted our subordinate status.

At times 40 elephants or more would silently surprise us, appearing behind or parallel to the Land Rover and shuffling silently along, paying us no attention. Tiny, adorable calves walked alongside their mothers. Other times, we sat on the roof rack, paralyzed by a barrage of mock charges. Angry mothers and testosterone-budding males inspired our desire to be as far away as possible.

Even solitary bulls made us jittery. One once approached the car steadily as if on a wrecking mission. Although we already sat inside, the night vision equipment lay exposed on the roof rack, beside cameras and our priceless identification photos. He stopped at the front bumper, and from there he dismembered us with his eyes. He shook his head, flapped his ears, and drew them back. His long grey trunk waved up and down, and back and forth, casing the car. We just sat, frozen in place, hoping that he didn't probe through Sonja's netted, open windows. Later we saw a similar scene, in the movie *Jurassic Park,* and compared T-Rex to our monstrous pachyderm.

Elephants frequently caused other problems as well. They chased away rhinos, destroyed rock cairns, and disarticulated my bone piles. After that it was impossible to judge distances for photos, and we couldn't rebuild the markers until the morning.

Other times elephants arrived frightened and nervous. They wouldn't drink immediately, undoubtedly because of us. The Etosha staff had told us not to worry, that in such situations it was all right for us to remain. The elephants would habituate. But we had our doubts. If the elephants were edgy and thirsty, we didn't want to harass them with our presence. Sometimes when groups kept circling the area but wouldn't come to drink, we abandoned our observations for that evening.

Lions too were impressive. In the northern Namib Desert they feared Hereros, Damaras, and Himbas. In those areas we had rarely worried about lions. Etosha was different. The park lions were not persecuted, but for unknown reasons the most aggressive ones were in our Zebra Pan study area.

A major problem was deciding whether to photograph rhinos when lions were near. There must have been times when we missed seeing lions, especially when clouds obscured the moon. One night we first saw two males and three females when they were only 30 meters from the Land Rover. They growled, gaped, and made mock charges. Eventually they walked on to drink, but then they returned, lying down only 25 meters away.

With a serious charge, they could have been on us in seconds. We debated whether to drop off the opposite side of the car and climb in. Just as we made the decision to go, Carol's eyes grew large. I turned my neck to see what she was looking at. Nineteen elephants stared at us from the Land Rover's other side, a mere 20 meters away. Our backs pressed tightly together as she faced the lions, I the elephants. We froze, our hearts racing, adrenaline exploding. Neither of us dared drop to the ground and possibly trigger an attack. We were helpless, intruders in a world dominated by elephants and lions.

To describe the majestic grace, the sheer power, of wild lions and elephants is like trying to explain the essence of an isolated coastline—the vastness, the sounds of crashing surf, the rich vibrant smells—to one who has never seen an ocean. At night and up close, elephants and lions unleashed primal feelings, mostly fear. But there was also a sense of beauty, awe, respect.

Back at Etosha Ecological Institute for 10 days every two months, those nights we'd sit at the floodlit water hole at Okaukuejo. There the rhinos were habituated to the tourists' commotion, chatter, and fumes. The walk from our caravan to the water hole was only three minutes. We didn't even need the moon; individual rhinos were well illuminated by the powerful floodlights. Some thought it was not macho enough to use these as study animals, claiming it wasn't a true African experience. We didn't. We were pragmatists. We wanted the data.

Doing observations at Okaukuejo did have drawbacks. During the winter, the prime tourist season, the hard wooden benches filled with people. To get a seat we had to arrive by 6 P.M. and remain steadfast until nearly midnight. Only infrequently would we bring Sonja. With all the distractions, she had trouble remaining silent. So we rotated duties, one of us staying at the caravan and carting dinner to the person doing obs. Temperatures that dropped to 40°F or lower meant hauling blankets or sleeping bags along with armloads of equipment.

Another difficulty was that we couldn't wander among the rhinos to take our horn measurement photos. Patience and luck were required. We recognized individuals, but it would take months for a particular animal to approach within 30 meters of the rock wall that separated humans from rhinos. Only then were the photos truly valuable.

On the other hand, watching the nighttime interactions at Okaukuejo was all the more interesting because the lights made them easier to see, and we didn't have to concentrate on our safety.

Our names for Okaukuejo animals represented Namibian peoples—*Bushman, Nama, Damara, Himba, Mafwe, Herero,* and *Owambo*. As we exhausted this supply of names, we added the names of local rivers, *Kwando* and *Kunene*.

People regularly asked us what we were doing and why. We gladly shared information, showing our photos, explaining our project, and describing rhino ecology, behavior, and conservation issues. But when rhinos were present we became serious. On occasion tourists would pester us until we let them look through the night vision equipment. We were hesitant, because a handling accident could have serious consequences, compromising our ability to gather data; but it was always awkward to refuse the requests. The tourists were there because of interest and, because it was difficult to refuse such enthusiasm, many left with a view stamped in their mind of a rhino at night magnified 13 times in green.

For fun we also gave a different type of name to some animals, names that greatly confused the tourists sitting next to us. *Himba,* a cow with a wheeze, known for ten years, had a male calf we called *Him*. A few others were recognizable because they were missing ears or parts of their tails. In a fit of juvenile humor, we named four animals *Exactly, Who, Her,* and *Really*. Often our friends the Gasaways sat with us and helped.

"Look, there's *Who,*" Kathy might say.

"No, that's *Really* or *Exactly,*" I'd respond, smiling to myself.

"*Exactly.*"

Sleep was precious. All-nighters had been events of the past, college memories, something that neither Cunningham nor I had ever expected to become a regular part of our repertoire. In the States we'd rarely managed a New Year's Eve vigil. We appreciated sleep. It's something craved during cross-country trips, that parents do without when their children are ill. In Etosha, although we often were not awake at sunrise, we fought sleep for up to 19 consecutive nights. It was not something that we could continue indefinitely. We became worn out.

Another cause of our general malaise might have been our allergies to grass. Or it might have been some bug that we picked up. Most likely, however, it was Sonja. She had her own schedule, rising with the sun, and like any normal child, she wanted playmates. "Play with me, chase me, feed me, paint, draw. . . ."

By midday my eyes would close, only to open sharply with a firm bonk on the head—Sonja. Cunningham would smile weakly and tell me that I had gotten 21 minutes more sleep than she.

One evening at the Okaukuejo water hole we watched three different female rhinos chase a solitary old lioness. At midnight an exhausted Cunningham walked back to the caravan, but I stayed, hoping for more rhinos. Six had already visited. By 1:30 A.M. I sat alone on the bench. Insects by the thousands pelted the floodlight, snacks for visiting bats. My feet were propped on the stone wall that separated me from dry grass and precious water. I struggled to stay awake.

Just as I was ready to admit exhaustion, the old lioness reappeared and padded toward the water. Then suddenly she turned and approached me. My eyes sprang wide open. Ten meters from the base of the wall, she crouched. My heart beat faster, the familiar adrenaline rush charging through me. Her tail straightened out and twitched back and forth.

Am I her next meal? What is she thinking? Lions do think, don't they? They must be more than a package of genes. Her eyes were large and yellow in the light, beautiful. All of a sudden the slanted wall seemed low and ineffective. *She can't scale this fence, can she? Nah. Otherwise, the management staff would have a higher or more lion-proof fence. Surely. Don't move, Joel, stay put. Don't breathe. Be part of this bench.*

The lioness turned and walked the 25 meters back to the water's edge. That was an experience I didn't want to repeat. And I wondered if a tourist would ever be eaten.

14

Namib Edge

[Joel]

F.. k the rhinos, save *the whites*," read the circular sticker.

"Do people really use these?" I had asked at a store soon after our arrival in Windhoek in 1991.

"No, not really. One shouldn't take it too seriously," replied the white attendant in perfect English.

He was wrong—something we wouldn't learn for another year.

A menagerie of stickers adorned the rear windows of our Land Rover. They proclaimed our ties to both Namibia and the conservation community. A large black rhino was the insignia for the Namibia Nature Foundation, the World Wildlife Fund had its panda, and a black-and-white gorilla set against green was Frankfurt Zoological Society's. A rhino against pink was Zimbabwe's. The South Africa–based Rhino and Elephant Foundation, Swaziland's Mkhaya reserve, and Save the Rhino Trust each had their own rhinos. Our display ended with Namibia's flag and leaping gazelles, the mark of the New York Zoological Society. Locals viewed the decals with a mixture of caution and approval. They suspected we were either a mobile panoply of environmental fanaticism or a research project with too many sponsors, perhaps both. The truth was that very few of those organizations were sponsors. We simply believed in them, and we and welcomed the frequent inspections or questions about our kaleidoscope.

Hence I wasn't surprised when a rough-looking fellow approached me outside Okaukuejo's little store, his voice muffled by the rumble of a noisy bus. Personal space wasn't an issue for him. Up close, he exhaled alcohol on my face. His eyes, set above a graying beard, exuded anger. He spoke in Afrikaans but, after seeing my blank look, he changed to English.

"Have you seen a bumper sticker, the one that says 'F..k the rhinos, save the whites'?"

I paused. *The man is serious.* My eyes moved slowly, deliberately, from worn leather boots and dark knee socks up khaki shorts and a burgeoning stomach to a ruddy face with a crooked nose. Inches apart, our eyes met.

"Yes, not very funny, is it?"

"More people should have them."

My jaw tightened. In disgust, I turned away. Bill Gasaway had just come out of the store and stood nearby, speechless.

* * *

A stack of literature crowded our caravan's tiny table. Some were photocopies from biologist Peter Erb's collection on African rhinos. Months earlier, we had visited his base on the plateau known locally as Waterberg Park. Although we didn't know him well, we had wanted to hear more about his work on kudus and rhinos. He had also been interested in our research and suggested that we duplicate whatever we wanted from his files.

We drove to Windhoek to make the copies. Because the facilities at Etosha Ecological Institute were primarily for government employees, outside researchers were rightfully expected to contribute their own supplies. Although we had our own paper for small amounts of copying, making duplicates of Peter's material was a major job.

Due to the political instability in so many other areas of the continent and the positive press over Namibia's recent break from South Africa, the Institute was inundated with external requests to work on wildlife projects in Etosha. Practical issues faced the Institute's research director, Malan Lindeque. He believed that studies must focus on problems of immediate consequence to Namibians, not esoteric problems only of interest to pure academicians. Malan also wanted to know about the prospective researchers' qualifications and political background, and especially whether they had leanings toward animal rights activism. South Africa and Zimbabwe were being targeted by groups from the West; according to Malan, the last thing Namibia needed was agitation by preservationists. Also it was critical that visitors not drain limited resources—office space, photocopying supplies, phones, or fax machines—or divert rangers, scouts, veterinarians, or biologists from their official duties.

At the core of these issues were concerns about conservation and sovereignty. Already several foreign projects had cut into ministry time and budgets. Malan simply wanted to be sure that any projects that might be approved did not go awry. Who could fault Malan's attitude?

Etosha biologists had fewer funds, less time, and poorer equipment than most foreign researchers. At one point the Institute ran out of photocopy paper, and the government's current budget didn't provide for more. Logistics, ill-defined job descriptions, and local emergencies always seemed to prevent turning research into reality. During the Institute's morning and afternoon tea breaks, tensions seemed to spill over. Conversations that changed abruptly from English to Afrikaans caused visiting researchers to cringe. It wasn't clear whether the shift was a deliberate attempt at exclusion or just a move to a more comfortable medium. Only later did we learn that outside researchers were often viewed as ecological imperialists, people who pirated Namibia's rich settings with their lust for scientific studies and who left little behind for overworked local biologists to study.

Malan was clearly in a difficult position. He represented the country at international meetings, he was supposed to carry out his own research program, and he made recommendations about the proposals and progress of foreign researchers. Bill Gasaway attempted to defuse the tension, organizing donations from British and American biologists working in Etosha, but government regulations prevented him from putting the money directly into Institute operations. Instead he contributed materials such as photocopy paper, ink, and a spotting scope.

Despite our infrequent visits to the Institute, we contributed a current subscription, with all back issues to the international journal *Conservation Biology*. We also agreed to leave behind photographic and radio-tracking supplies.

Among the proposals park biologists discussed with us was a study of the black-faced impala. These graceful antelopes with dark blazes lining their faces, which have been described as "living fossils," differ from the common impala, which is distributed widely from Kenya to South Africa in woodland and mixed savannas. In Namibia, black-faced impalas were once found throughout moister areas of the dry northwest, but these Kaokoveld populations had been virtually exterminated. To prevent the total extinction of what was thought to be a distinctive race, animals had been introduced into Etosha. However, little was really known about the genetics, population status, or ecology of the black-faced impala.

Carol and I were delighted when this new project was sanctioned. The would-be researchers were friends, Drs. Wendy Green and Aron Rothstein, who had studied North American bison and were research associates at the University of Nevada. They'd visited the year before and could soon begin their project. They would be based far to the west of Okaukuejo.

The proposals of other scientists fared less well. A study of elephants in the desert by Ian Douglas-Hamilton was quashed. A Swedish scientist hoping to investigate whether antelopes with disfigured horns were more susceptible to predation was denied access to the several hundred skulls in the Institute's collection. An inquiry about Etosha's butterflies didn't have tangible benefits. And the proposed continuation of a dik-dik study on ecology and reproduction by a zoologist at University of Cambridge was not allowed.

We felt lucky that Malan, Save the Rhino Trust, and Eugene Joubert had all been supportive in the developing stages of our work. Studying controversial species, and particularly a topic like dehorning, could have drawbacks, but at least our project was approved.

*   *   *

It was cloudless, dusty, and hot when we arrived in Sesfontein to retrieve Archie and begin another round of data collection in the desert. Ruppell's parrots passed overhead, offering a fleeting glimpse of grey heads, thick bills, and feathers with bits of blue and yellow. Near Archie's hut, goats bleated and then tripped aside. A more grown-up Sonja climbed from the car; now two and a half years old, her vocabulary had improved, and she no longer wore nappies. She even tried talking with the local kids, but they were still more interested in touching her hair and skin. Archie ducked his head through the low door to his home and began pulling his gear into the sunshine.

His parents happened to be in Sesfontein, and we exchanged greetings, ours in English, theirs in Afrikaans and Damara. We all laughed and understood each other perfectly. We gave a pocket knife to his father, a scarf to his mother, and sunglasses to each. They offered dates and corn grown at their farm more than 160 kilometers away. Archie would be with us for a month or more before returning. He waved goodbye to his family and squeezed into the back seat.

Afternoon winds roared through the open windows, and Archie's quiet voice was that much more difficult to hear. He told us about a woman who had been lost somewhere in the desert south of Sesfontein—a 58-year-old Nama. She'd been missing for two days and two nights, despite people searching for her.

"She disappeared. When they found fresh lion spoor, everyone was very worried."

The police heard about Archie's tracking skills and drove more than eight hours to retrieve him from Hobatere. It was a desperate gamble, but few options remained. Archie was the last hope for this woman.

A faint, wind-blurred human spoor remained. Because of their small size, Archie assumed the footprints were female. The tracks twisted through ravines, over rocks, and into a canyon. All concentration and pursuit, he followed her trail for nearly 20 kilometers. At last he found her, face down in the midday sun but still breathing. She would be all right. The police thanked Archie and returned him to Hobatere.

"Just another day at the office. Right, Archie?"

We talked sporadically, shouting in the rushing wind. Dust trailed behind the Land Rover as landscapes shot past. We described the Etosha lions and told Archie about the recent rhino poachings in Angola just north of the Caprivi Strip. He gestured abruptly to one side, pointing to gray dots on a distant acacia-covered slope. Two elephants were feeding. I craned my neck to the left and looked into

Cunningham's blue gray eyes. We smiled at each other. It was good to be back in the desert, and with Archie again.

We decided to go to Khorixas first. I wanted to find three skulls that the ministry had, and Archie needed to cash the last three checks we'd given him.

As luck would have it, the bank hadn't been open for days. People milled around outside locked doors. Others waited patiently in a line that filed past a wire gate and along green manicured hedges. No sign explained why the doors were closed, and no one seemed to know when the bank would reopen. Archie learned that thieves had robbed a vehicle transporting money, and there had been a shootout. Although the would-be robbers had been caught, no one had yet decided that it was safe to drive another vehicle with money from Windhoek. Archie's uncashed checks would have to wait.

Near the bank, two donkeys were tied to a post. Their heads jerked back and eyes rolled nervously when Sonja and I approached for a closer look. Clicking and smacking in Damara, the owner told Archie that the animals were afraid because they didn't see many people with white skin. He seemed surprised but pleased that we were interested.

Just before leaving town we located the skulls. Two were nestled against a wall in Rudi and Blythe Loutit's backyard for safekeeping. Another sat in a cardboard box packed away in the garage of a different ministry worker. Using published tooth diagrams, we estimated the animals' ages and took skull measures, information that would later suggest relationships with the horn sizes of these dead rhinos. Slowly our perseverance was paying off. Our sample size of skulls of known sex and estimated age had grown to 19.

At the petrol station, surrounded by a crowd of small boys and girls, we filled half a dozen jerry cans and our three petrol tanks. A Peace Corps worker happened to walk by and told us that leftover American rations from the Gulf War in Iran had been distributed in Namibia, some even in Khorixas. Avid and hungry eyes watched when, after months of waiting, the precious goods were opened. *Gherkins.*

"Just what poor Hereros and Damaras needed, packaged pickles," remarked a local.

The drought intensified. The 1992 rainy season had been a bust, and now more than half of the 16 spots where rhinos had drunk in 1990 were already dry. Perhaps rhinos would be easier to find. With water relatively concentrated, rhinos might not move more than 15–20 kilometers from it, and then they'd be within an 800–1,300-square-kilometer area. We were optimistic.

Eight days passed, and so did our confidence. We couldn't find a rhino and were perplexed by human footprints and tire tracks that spread across 4,000 square kilometers of raw desert. In these remote areas Carol and I had never seen so much human sign. Neither had Archie.

None of us suspected this to be the work of poachers. People who shoot rhinos, at least in this part of Namibia, are secretive. They work on foot or with donkeys, not with vehicles and hard-soled hiking boots. To continue our search in areas recently trampled by a virtual army—teams of vehicles and trackers—seemed fruitless.

Leaving the area and driving south, we speculated about the sources of disturbance. I recalled a rumor I'd heard months earlier. The ministry, in consort with Save the Rhino Trust, might launch a massive project to census every rhino in the northern Namib Desert. Perhaps it was no rumor. Maybe the effort had begun.

The more we talked, the more we convinced ourselves that this must be the case. How would a novice know to search virtually every nook and cranny of the desert? It was clear why Archie knew nothing about the possible operation. He was a security risk. But he couldn't understand why we knew nothing of the census.

We'd try to find rhinos in other areas. As we passed abandoned farms, homes of Damaras who had moved on, flea-bitten donkeys appeared against a blue skyline, and water trickled from the ground into a trough. A man with four boys stood at a borehole. I shot a quick smile at them, shook hands with the father, and revealed that I couldn't speak Damara. Archie helped. He explained that we were from America and that he helped us track rhinos. I recognized the Damara word !Nabis, meaning rhino.

The sun was hot. I motioned to see if the group wanted to share in some juices as we talked.

"Yes," came the reply from a young man, whose age I guessed to be 15. "That would be great."

Shocked, I smiled, thinking *such good English, not the words of peasant farmers. Who are these people?*

The young man had been learning English in Khorixas for five years. His father owned a farm 20 kilometers from us. Soon sheep and goats would overrun this lovely desert spot.

I returned to the Land Rover where Carol, Soni, and a visitor along for a short ride waited.

"Juices! You're making a mistake. Giving them juice is like saying it's okay to stay," came the visitor's shocked response.

"Better to be friendly," I pointed out. "Besides, we aren't going to change what they are doing or how they think by not being nice."

"How long will they stay?"

Not knowing, I shrugged.

"What, you didn't ask?!"

Finally annoyed, I muttered, "It's their land, for Christ's sake, communal land."

"The rhinos have been here a lot longer."

"Maybe so, but that's not the point. It's their land, and it's been occupied by people a hell of a lot longer than we whites have been around." My message fell on deaf ears.

"Look, calm down. Maybe they can even tell us something about rhinos or hyenas in the area," suggested Carol.

"Why not ask them to move in permanently!" fired our guest.

His view was like that of many Europeans and Americans. Wild areas were for animals, not people. It mattered little whether people had been there for millennia.

We reached Doros Crater. Where foliage had once formed an intriguing patchwork of green and yellow, scattered among schist and volcanic rock, the landscapes now lay bare. There was no grass. Bushes were brown and dormant. Without food, animals would starve or migrate. Even desert-adapted species like ostriches or gemsbok, animals that go weeks without surface water, had vanished.

For three days we inched along, driving 80, 90, and 140 kilometers a day as we searched north and west of the crater. Around the fire at night we talked about where to try next. Archie's intuition was excellent, but occasionally I made the decision. At this point who decided didn't make much of a difference.

On the fourth day we returned to camp to find that old friends had stopped by. Their calling card was curious, a mixture of plastic and other trash strewn on the dirt. The pied crows had won another round, raiding our cooking supplies and tearing apart bags of garbage. Before leaving camp the next morning we piled Sonja's red tub and a heavy tarp onto the trash heap, weighting down the whole mess with rocks. Although this day too passed without a rhino, we discovered a small reward at camp. The tarp and the red tub were exactly as we'd left them. Even such insignificant victories were sweet. But that evening the short-wave radio reminded us of just how trivial our troubles were. Ten more people had died violently outside Johannesburg; 9 foreigners had been killed in Angola.

The next morning we all agreed to search for rhinos elsewhere and packed hurriedly to leave. Soon we were churning through deep sand. I drove faster than normal to keep the necessary momentum. When the riverbed abruptly cut to the left I accelerated, but I couldn't turn fast enough, and the Land Rover smashed into two stout boulders. Sonja tumbled into the dashboard, screaming. Archie, on top, catapulted forward but managed to hold onto the rack. We were all unhurt and amazed that, other than a small dent in the bulbar, there was no damage. The metal bar on the front of the vehicle had been welded on for extra protection and the Land Rover was fine, a true tank. Seventy kilometers later, we wearily set up yet another temporary camp.

None of us had been to this canyon south of the Ugab River. According to the map we were in Brandberg National Monument. Now we truly believed our luck would change. On the way we had seen the spoor of at least five different rhinos.

As Archie and I worked the Ugab and its brutal gorges, my feet developed painful blisters, but not before some success. We found *Suzette,* a young horned female, daughter of the recently dehorned *Susie.* Later, in dense shrubbery, we managed to climb a tree where, over Archie's protests, my imitation of baby rhino calls enticed *Mad Max* to within 5 meters. His thickened horn base showed impressive regrowth. We were all anxious to have Carol print photos so that we could estimate how much the horns had grown since dehorning.

I stripped away sweaty socks, ignoring their rank odor. Carol frowned. She looked at my bruised and blistered feet, and with great sympathy, muttered, "Oh brother, I can't believe what a pansy you are." Then she dropped a few bandaids on my lap, hoping, I'm sure, that I wouldn't ask her to track. Her recent experience with the charging female had only strengthened her fears, but in that situation she could see. In this part of the river, the dense thickets were horrifying, not only to her but to Archie and me. Unhappily she agreed to track again.

On her first day out, she and Archie discovered soft fecal patties. The rhino was close but hidden completely. Brush taller than humans and mixed with acacia and thorns eight centimeters long filled the canyon's bottom. Archie climbed the canyon's walls for a better look, peering slowly into the thick vegetation for movement, a twitch, anything.

"I was so tense that my stomach hurt. I just don't have the temperament for this," said Carol to me later. "At one point we were walking on a path that tunneled between walls of brush, and we were sure a rhino was right next to us. Branches started snapping and crackling loudly. Archie whirled around, and when I saw his eyes get huge and scared and his arms begin to pump, I turned around and ran as fast as I could. But there was nowhere to go. Thorns and branches were everywhere. Then I noticed it was quiet behind me. Archie had disappeared. I was alone, thinking *I've got to get off this trail or the rhino will be right on top of me.*"

"I didn't care about the thorns any more and just jumped into the bushes, and after two or three steps I was completely stuck. It was pathetic, I was still right next to the trail and terrified. It was only later that I saw blood dripping down my legs. Finally, I noticed the total silence. No rhino was coming down the path. My voice wavered when I called out, 'Archie, Archie?' "

" 'Here Carol.' His voice came from no more than 20 meters away."

As Carol's story ended, I saw fear in her face. But she persevered and the next day photographed *Suzette's* mother, *Susie*, at close range.

After only two days of reading, playing, and sizzling with Sonja back at the Land Rover, my blisters had healed remarkably. I was ready to track again. This time we found *Skellum*. Two days later we got a different male, *Kallie*. Our data base grew steadily. We had information on regrowth rates for 65 percent of the dehorned rhinos. Only three Doros Crater animals remained.

The next animal we sought escaped into a deep canyon, running in the direction of the Land Rover and Cunningham, still miles away.

At top of the hour we called her on a portable radio. The transmission was scrambled, but some of her words were clear.

"Seven men . . . I don't think . . ."

"Repeat, Please repeat," I yelled into the radio.

Left: Government workers sawing the horns from a tranquilized male rhino.
Right: Remaining stubs in the process of being clipped.

Left: Himba charges a lioness at dusk in Etosha. *Right:* Sonja displays a horn
collected during dehorning operation. Dr. Pete Morkel, shaded, and Rudy
Loutit in background.

Black rhinos are nature's tank. They can run 50 kilometers per hour.

Composite illustrating individual differences. *Left:* variations in places where ears are torn. *Right:* disparities in horn size and shape.

Two young aardwolves in a sandy den.

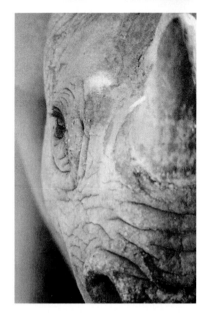

Dehorned rhino with two years of horn regrowth

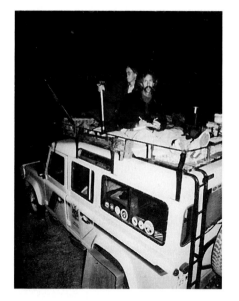 

Left: Our nightly set-up. Carol scans with the night vision scope.
Right: Goat herds near Sesfontein.

Left: A Land Rover struggles in the canyons between Doros Crater and the Ugab River. *Right:* Bushman children enjoying their playtime with a very unhappy Sonja.

Elephants and
giraffes in
Etosha.

A springbok
herd moving
through a dusty
haze at sunset.

Early morning,
smoke, and
coffee.

Somalia patrolling his territory as zebras look on.

Tina and *Tiny* with two springbok and an abundance of Euphorbia.

Desert rhino walking away from government workers after dehorning. Photo from helicopter.

Sonja with field notebook and bullet-ridden elephant skulls, Wereldsend.

A spotted hyena, part of the Zebra Pan clan.

Malawi (female, *left*) and *Sudan* (male, *right*) with unsure intentions. Note *Malawi's* young daughter to her left.

The aftermath of poaching: rhino skulls at Wereldsend.

Engravings on a rock slab at Twyfelfontein.

Sets of tracks from three species: gemsbok (*left*), rhino (*middle*), and elephant, (*right*).

Sonja and Joel enjoy pool from a sudden downpour.

Carol reading to Sonja on the top of the Land Rover before night observations at Calcrete Basin.

Left, Carol and Sonja roasting at the bottom of San Cave Canyon, where ground temperatures reached more than 150°F. *Above,* A dead rhino showing signs of anthrax. Note bloody discharge from nose.

Carol peering from the tower
at Calcrete Basin. Note herd
of elephants.

Sonja shows Archie her dolls, Marianne
and Patch.

A Zebra Pan elephant swings his
trunk at a timid rhino.

Archie and Joel struggle to free a
springbok.

One of Etosha's many
dung beetles.

A yellow-billed hornbill. *(Photo by
J.A. Phillips)*

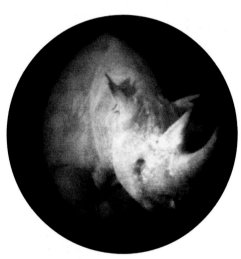

Africa photographed through night
vision scope.

Namibia contains Africa's largest
cheetah population.

A small band of mountain zebras near the Uniab River.

Emperor moth with juice container.

Playing ball on the airstrip.

A charging rhino (*Kokerboom*) in the Namib Desert. Carol is at the base of the tree. Archie is in it.

Kokerboom approaches Carol, still struggling toward safety.

Carol sits atop the mired Land Rover.

Baking bread in the cast iron pot.

Himbas, Carol, and Soni in the northern Kaokoveld.

Sandy two years after dehorning. *(Photo by Archie Gawuseb)*

Sesfontein during a windstorm.

Archie in front of his home, strapping on supplies as his daughters and Carol help. Note the fender to the right of the rear tire crumpled by an angry rhino.

Warthog.

Camp invaders.

Elephants, one dead, one alive, and vultures.

Archie and Joel and the Kunene
River. In the background: Angola.

A Zimbabwean memorial in Hwange
National Park to rangers who died trying to
protect rhinos from poaching.

Two young Himbas.

Measuring horns in a vault. (*Photo by Franz Camenzind*)

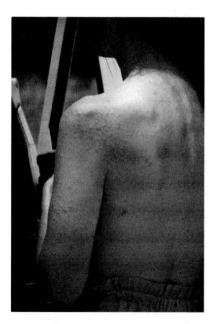

Some of the mosquito bites from our night along Kavango River. Carol's good side.

Kipsigi aggressively approaches an elephant bull at Okaukuejo.

Archie 75 meters away from *Kai Kams* and her calf *Kaiman,* in the gravel barrens.

The lions and cubs just before we were charged.

Salvaging our supplies after the Land Rover fire near Doros Crater.

"Seven men . . . within 500 meters . . . I don't know if they saw me . . . spread out . . . coursing."

She was unsettled. Me too.

"Were they carrying weapons? Did they see you? Were they barefoot?" Transmission breaks.

"I don't"—transmission breaks—"I don't know."

"Which, what don't you know?"

The transmission improved and I heard.

"I couldn't tell what was in their hands. They carried something that looked like it was wrapped."

I repeated, "Did they see you?"

"I don't know, I was lucky to get a count."

I asked if they could be miners, farmers, or tourists.

Archie interrupted, "No."

"Why not?" I said.

"Joel, tourists would have a vehicle and hiking boots. They would come over and talk." He thought a farmer would send boys, and not so many men, to look for missing goats or cows.

Archie added that if they were miners, they could be interested in only one thing—rhinos. Downriver there might still be ostriches and kudus for food but, where we were, there were only rhinos. Archie and I also knew that a rhino had just fled in the direction of Carol and, now, the men. We raced to find her.

At the place we discussed—southwest of the Salvadora thickets—she was waiting. The three of us discussed our next step. In case there were problems, Archie and I wore guns. We grabbed extra ammunition. Carol's face looked pained. We looked at each other. I managed to wink. She smiled weakly. I departed to follow Archie.

As we tracked back into the canyon, thoughts swirled in my mind. *What am I doing? Why am I carrying a gun? I'm not a warden. I don't have the training. Wild animals are one thing, people are another, especially people intent on killing rhinos.*

By noon the canyon heat had swelled to more than 104°F. Navigating each thicket was unsettling. My fears were balanced by knowing that the human tracks were sporadic. Neither of us was certain why. Archie could track an ant across a golf course. Perhaps the men had made special efforts to avoid detection. Despite the tension, Archie fascinated me. His lean body bobbed, turned, and weaved. His head jutted forward and sideways. He listened. I tried but just didn't have the same skill.

At 1 P.M. we ate biltong and canned peaches, then guzzled the warm syrup in the shade of a small alcove. Lunch had been the same for four days. I didn't know what to think about the men and remembered memorials in Zimbabwe, tributes to rangers who died while trying to protect Zambezi Valley rhinos. I closed my eyes and dozed. . . .

Archie jumped up as strangers came into view. I counted men—four, five, six. Where was the seventh? Was Carol's count wrong?

My heart sank when I saw that two of the men were armed. Archie undid his weapon. I pressed my back against the rock wall and reached silently for my gun.

We shouldn't be here. This can't be happening.

It was too quiet, and real. The sun pounded. Sweat poured from my head.

The men advanced—40 meters, 30 meters, 20. Archie stared, his arms low. I continued sitting and heard pebbles falling from above.

The seventh man must be up there.

I remembered the final scene in the movie Butch Cassidy and the Sundance Kid*—outnumbered, the duo died in a shootout.*

One of the men came closer and Archie said something in Afrikaans. I didn't understand. Arguing began, shouts escalating. I wanted to be somewhere else.

Flies, why are there so many flies?

I blinked and wiped sweat from eyes. In that millisecond, three shots rang out. Two men writhed on the ground. I looked at Archie. Blood gushed from his chest. I pulled my gun. My eyes burned.

The sound of a snapping branch and a foot nudging my leg woke me.

"Joel, let's go, are you ready?" said Archie.

Little did I know that my dream could have been Carol's reality. As she drove to our rendezvous site, a bakkie appeared in front of her, blocking the rocky spoor. Two men were in the open back of the truck, another up front with the driver. There was no way around it. Sonja sat strapped in her seat. *Were these the men that she had seen?*

As Cunningham approached, the bakkie slipped into gear and moved forward. She didn't know whether to stop or follow. She kept going. The two vehicles, 20 meters apart, crawled along an isolated two-track in the Namib Desert. A dark hand reached out a window and adjusted a mirror. The men in the back of the bakkie turned and stared, straining for a better look at the Land Rover. Carol memorized essential features of the vehicle—its battered body, yellow doors, white top, dull green frame. For three kilometers she followed, slowly and deliberately.

The bakkie finally stopped just beyond a forked junction. The men climbed out, and the large, rotund driver straddled the rocky spoor. Carol slowed. But just as she reached the junction, she pulled to the right, waving an acknowledging arm, and accelerated toward our arranged meeting place, some 60 kilometers away.

Archie and I took the shortcut, through the canyon and over a mountainous spine. We arrived in two hours, knowing nothing of Cunningham's predicament. We never saw the rhino or the men.

At the Land Rover, we discussed the day's events. Archie told Carol that she did the right thing—not stopping. We decided that we'd call Rudi Loutit, the chief conservator of the area, and report the affair. Although it would still be days away, at least he'd know what we'd seen. It was the last we ever knew of these men.

Our search for more dehorned animals continued. During our rhinoless days, with Archie "spotting" from the rack, Cunningham and I talked about life, relationships, and family, and reflected on our project. We couldn't arrive at a reasonable explanation for why we hadn't been able to find rhinos after our discovery of human tracks and spoor in the Wereldsend area. One possibility was that rhinos had given up using those areas and moved elsewhere. That didn't seem plausible, as Archie, Garth Owen-Smith, and others felt that rhinos were nonmigratory. Another explanation was that rhinos abandoned areas once they encountered humans—at least temporarily. If, in fact, rhinos were fleeing from humans, then tourism might not be the panacea that was so widely touted in conservation circles. The presence of large numbers of humans in desert areas might cause rhinos to leave, burning off precious energy reserves, abandoning watering sites, or giving up access to valuable food.

We already knew that the rhinos around Okaukuejo were habituated to humans. So were the rhinos in Ngorongoro Crater and in isolated reserves in Kenya, Zimbabwe, and South Africa. But if rhinos were not regularly exposed to people, habituation was impossible. The consequences of human disturbance in places like the Namib Desert, where rhinos live at the fringe of their range, conditions are marginal, and humans are rare, might be negative. If rhinos were regularly displaced from their preferred feeding and watering areas, they would be forced into areas that were even more marginal. At some point reproduction might be compromised. Gradually, through attrition alone, the animals might perish. It was all speculation, but we noted that a key aspect of rhino conservation in this harsh environment was understanding the distribution of rhinos in relation to humans, tourists included.

Our eighteenth day on this field mission began like all the others. I woke to the snapping of twigs and a glowing eastern sky. Archie had started the morning campfire. Carol curled inside the sleeping bag. With each of my descending steps, the Land Rover rocked and the ladder creaked.

"Whoooooosh." Archie's match loudly met petrol. Water for coffee and tea steamed away. I peeked at Sonja. Ted (her teddy bear) was cuddled in her delicate arms; both were tucked in in the tattered tent. Engrossed in their sleep, mother and daughter missed the sweet smells and soft light of early morning. We were all smelly, filthy, and low on water and petrol. We'd give one more day to rhinos before refueling in a town.

At 6:55 A.M. we pushed off. At four o'clock that afternoon the odometer clocked our ninety-seventh kilometer. We were somewhere south of the Ugab River and exhausted by the canyon's heat. I almost hoped that we wouldn't find fresh spoor. A reflection far off in the next tributary caught our attention—a vehicle. As we approached, we could see it was new, a 1992 24-valve Land Cruiser. The bonnet was up, and four travelers beamed at us. Archie sat in his familiar spot, the old mattress on the roof rack. Carol and I walked over to them. Greetings were exchanged. The new 180,000R ($60,000-dollar) vehicle's problem had been solved just minutes before we arrived. These were two middle-aged couples on vacation from Cape Town; they had driven inland from the coast when their car failed.

One of the men pushed cold beers into our hands. We declined. Archie didn't drink, and neither Carol nor I would be able to follow a rhino if we downed the beers. We explained who we were.

"There are rhinos in an area like this?" one exclaimed.

At that point another admitted that, had they known, they probably wouldn't have wandered so freely through the thick Phragmites reeds.

As we left, Cunningham and I looked in the mirror. Her T-shirt, covered with stains, was stiff with dirt and sweat. She'd only worn it for four days. I was worse— no shirt, a scruffy beard, unruly hair. Sweat cemented with suntan oil formed a protective skin shield. Archie looked like an old man, his black hair gray with fine silt. Our daily allotment of a cup of water each wasn't designed to keep us presentable.

We stopped at Wereldsend to see whether anyone knew about the mysterious tracks to the north. To our delight, Garth and Margie were there. Eight months had passed since we'd seen them. They were often out working, but they knew about the Kaokoveld rhino census and confirmed that a survey was already in progress. The tracks that blanketed the landscape had been made by government and Save the Rhino Trust employees. Our suspicions verified, our disappointment returned. If we had been informed, we wouldn't have wasted nearly two weeks in an area searching for rhinos that we had little chance of finding.

We told Garth our ideas about human disturbance and local site abandonment and about the vulnerability of males and females to poachers. He was skeptical, but that was normal. Garth was a cynic, Margie more so. We all delighted in vigorous discussion, debating each point and reciting one anecdote after another to support or refute each idea. In the end, Garth conceded that the scenarios might have some merit, hoping that by the time we returned to Namibia from the United States in 1993 analyses would be complete and an objective examination of the ideas would be possible.

That evening Garth showed us his records on rhinos—files that consisted of locations, births, and deaths, and meticulous drawings—information accumulated during the last eight years. Among these were photos we had sent. We told him where we'd seen the animals, updating his files with our records. When he told us that he'd be willing to share information, I didn't quite understand the gesture. Finally, I grasped it. He meant that we could photocopy his files. My mouth fell open.

So did his, when I uttered, "We'll do it to tomorrow."

He didn't understand our enthusiasm about driving 800 kilometers round-trip to make copies. I explained.

"We've worked for weeks for a few data points, and what you're offering took years. Driving 800 kilometers is a good deal, I'd say."

Garth understood.

We followed the Veterinary Cordon Fence part of the way, its rigid cables and fenceposts clearing the innocent landscape. In the distance, a tawny body with delicate legs and a dark band hung from the fence. A springbok's front leg was so entwined that her hind legs were hoisted nearly a third of a meter above the ground. We drove up to investigate. She was still alive. Wire pinged and pitched as she kicked violently, bleating wildly. Archie and I dashed toward her. If we couldn't free her quickly, or if her leg was broken, we'd have to shoot her.

I seized her horns while Archie held her legs tightly. Neither of us wanted to get hit by sharp hooves. Inside the car Soni stared, captivated. Sixty seconds later the animal raced toward freedom, up over a hill and away from us.

Only once before had we seen an entangled springbok. In Etosha two males had locked horns in combat and lay on the ground, unable to extricate themselves from each other. Vultures waited nearby.

We continued our search for rhinos. More than a kilometer away a lone springbok, a female, stood on a gravel plain.

15

Buried in Sand

[Carol]

Not much lives on sand when air temperatures hit 122°F. The Namib's raw desert of sand and sun requires special ecological adaptations. One local species coping with these torrid living conditions is the Namib ant, known also as *Ocymyrmrx robustior.* Biologist Alan Marsh found that these long-legged ants, which don't become active until the ground reaches 87°F, tolerate sands heated to 151°F. Leaving subterranean nests to search for food, they cool themselves by pausing on grass stems, where it may be a remarkable 36°F cooler.

Our immediate problem, however, wasn't fiery sand, it was visibility.

We had crossed into a new phase of desert. Sand lay not only in dry rivers but in dunes that reshaped landscapes. Torrents of sand pelted the Land Rover, obliterating midday vistas. The skies darkened as the horizontal grains flew faster and faster. Headlights on, we plowed forward at 10 kilometers an hour.

Within minutes the storm ended. Soon afterward, a beige bakkie shot past but then screeched to a stop. Out stepped Rudi Loutit. Dark-haired and in his forties, Rudi had devoted twenty years to working these remote deserts. The frustrations had to be indescribable, the victories rare. But Rudi was totally committed and always ready for a good fight. He was also unpredictable. On this day we were lucky; he was talkative and friendly.

He told us about the latest schemes to mine the desert. Norway planned to bore for underground water, America to excavate for minerals. The sites chosen were the Ugab River and the lands within Brandberg National Monument.

Under the right circumstances mining makes sense, but in the unadulterated Namib it didn't, at least not to us. Archie understood both sides, ours and those of the local Damaras. Extraction industries can provide large benefits, sometimes quickly. At the same time, few local people were profiting from the wildlife on their lands. During 1992, moneys derived from tourists visiting the vast deserts went primarily to white-owned concessions, not to the farmers or local landowners. Because hunting was illegal and wildlife had little immediate value, the local peoples were pleased by the prospect of a new mine. In westernized societies people have greater opportunities to move and to be educated, and both leisure time and disposable income are widespread; rural Namibians have no such luxuries. For them, arguing for mining is easy, because mining equals jobs. Long-term gain is not an issue when weighed against the urgent need of immediate money.

But there was another point, one that Archie raised regularly in our discussions about conservation. With wildlife, he argued, tourists will come, now and in the future. Without wildlife, in a land full of roads and mining scars, there is no beauty and no reason for people to come. We liked Archie's perspective and hoped that he'd convey it to people in Sesfontein, Khorixas, and throughout the Kaokoveld. Archie's views had been shaped by Rudi as well as by local conservation groups. We wished more of the world shared similar beliefs, but our cynicism had been fueled by too much exposure to international corporations operating without conscience.

Cigarettes were everywhere. Archie had no idea that smoking was dangerous. In the United States alone, the costs of smoking-related health problems were estimated by the Center for Disease Control at $50 billion annually. Tobacco products must be labeled as potentially hazardous there, but not when marketed in southern Africa or many other areas of the world. Archie had a difficult time believing our claims until we showed him American magazines with health warnings on tobacco products. Only then did he seem to think we had a point, that economic profits drive policies. But he didn't give up smoking.

On our way to pick up Archie this trip, we finally visited the "Impala people." Aron (Rothstein) and Wendy (Green), our friends from Nevada, who were just getting their project launched. Their small caravan was tucked below a shady tree at Hobatere; among hornbills and lion roars, we joined them for dinner. It was good to catch up.

The next morning, as JB reversed the car, we heard a loud dull crunch, followed by a few muttered obscenities. I knew there was going to be a problem.

"What was that sound, Daaadddy?" asked Sonja.

Reluctantly, Joel replied, "Remember when Daddy accidentally backed into the tent? Well, Daddy just ran over the red tub."

"I want my tub, Daaadddy." Her face reddened to match the tub as tears streamed down her little cheeks.

I wasn't sure where we were. During the rocking drive, I had entertained Sonja while Archie watched out the passenger window. The Land Rover tilted along jagged, tire-puncturing ridgelines somewhere between the plains near Guantajab and the deep tributaries leading to the Ugab. At last Archie found a rhino spoor, but none of us was sure it was fresh. The winds were strong and could quickly obliterate the crispest prints. We had no idea when the animal had passed.

At 2 P.M. we stopped. JB and Archie left to poke around over a couple of ridgelines. Because we still needed to find a good spot for camp, they didn't plan to be gone long. Once we decided on a site, supplies needed to be downloaded, petrol and water jugs stored out of the sun, wood gathered, and tents pitched.

By 6:00 I assumed they'd found a rhino. Why else would they still be gone? They didn't have a radio or food, and I wasn't even sure if they carried water. Their short foray had lasted four hours. Sonja ate tuna over crackers, and we shared an apple.

At 8:00 it was almost dark. Settling Sonja in her back-seat bed, I ran the engine and switched on the headlights, aiming them at the crest of the next hill. Maybe Joel and Archie would see the light. I also honked the horn every five minutes or so. If they couldn't see, maybe the horn would reach them. I saw nothing. The rocky hills seemed uninhabited. Only the mechanical humming disturbed the silence.

By 10:00 I was frantic. *Maybe one of them broke a leg.* Or worse: *Maybe someone was gored. What should I do?* I couldn't think of anything else. This had never happened before; we'd never even made any contingency plans. I couldn't drive out to look for them. After disappearing beyond the near ridges, they could have hiked in any direction. At 11:30 I cut the lights and ignition, admitting to myself that they wouldn't be coming back, at least not tonight.

Cocooned in a sleeping bag on a tarp, I tried to rest. At 12:30 A.M. I woke with a start and listened to the quiet. A cool breeze blew my hair. I worried and couldn't sleep. I checked on Sonja and shined the flashlight into darkness. Distant bushes glowed back. I thought of scorpions and curled more tightly into my bag.

Before first light I woke again. Without Joel beside me I felt naked, vulnerable. Before the sun hit, I woke Sonja.

"C'mon sweetie, let's go for a hike. We'll have a picnic breakfast on top of that big hill. See it?"

I grabbed a little food, water, binoculars, a spotting scope. For Sonja there was a toy and book. With her on my back, I walked to the tallest ridge I could see and set up the spotting scope, depressed and exhausted. Sonja was entertained by the change in routine, and I tried not to show how tense I was. Taking care of her gave me some relief, but it was sporadic. I was afraid for both Joel and Archie.

Where are they? Both took more risks than I did, which wasn't saying much, but neither was careless. They knew rhinos. They regularly devised a plan before moving in for photographs, and they worked well together. When necessary, they shielded each other by distracting rhinos with noises. I kept hoping it was only a broken bone and that the other one was uninjured and just waiting until morning before seeking help.

There was still no sign by 8 A.M. I considered driving the 160 kilometers to Khorixas to organize a search, but if JB or Archie showed up while I was gone and the other was hurt, there wouldn't even be a vehicle for transport. I could waste days, increasing the danger if someone needed help right away. We had plenty of water and food. Soni and I would just sit for awhile.

Already the rocks were heating up. The only shade in the small valley below us was cast by the Land Rover. I peered through the spotting scope. Doros Crater jutted above a band of grazing mountain zebras. There were ostriches and spring-bok on the distant gravel plains.

At nine, movements behind me grabbed my attention. Several small birds, probably gray-backed finch larks, had landed. I showed Sonja more pictures.

Half an hour later I looked up and saw two tiny figures 50 meters from the Land Rover. *No, I'm just hallucinating,* I told myself. But this time I was right. Joel and Archie were trudging toward the car. My heart beat wildly as I stood on the crest, waving. Joel waved back. Neither seemed to be hurt.

"Look Soni, there's Daddy, Daddy and Archie." We hugged each other and then inched down the mountain and back to the Land Rover.

JB and Archie sat in the Land Rover's vanishing shade. By the time I got there, four juice containers lay empty. Both men shoveled food into their mouths. Sonja ran the last 30 meters and gave Joel a hug. Then it was my turn. Joel's taut, sweaty body felt good.

Archie collapsed in the remaining shade. Joel slithered under the car. Both were asleep when they hit the ground, but not before telling us what had happened.

At around 4 P.M., they had dropped their packs, thinking they were one canyon over from the car. But then they couldn't find us. "It all looked the same," said Archie, "every canyon, every hill. We didn't know where we were."

"Everything was in our packs, the emergency supplies, matches, my knife, even spare snacks and water. All I grabbed was my T-shirt, sunglasses, my hat and binoculars. Archie had his smokes. We kept thinking we were close to the car and we'd pop over the next ridge to find you there," said JB.

At dusk they finally accepted the fact that they were lost. Although it was quickly becoming very dark, Archie was anxious to keep moving. Joel felt it was best to stay where they were until morning, but it took time to convince Archie. They stopped walking at 9:30 P.M. Cool coastal winds were blowing across the raw desert, and without shelter and extra clothes, they were getting colder and colder. The temperature had probably dropped from 90°F to 60°F. Both were in shorts. Archie had only three matches. The last was successful in lighting a bush for warmth, but it soon burned out.

They dug shallow pits, 30 centimeters deep and 2 meters long, and lay in them, covering themselves with sand. With their faces exposed, they worried about a rhino or hyena coming along. "Faces might make nice morsels," JB had joked. As the weight of the sand pressed on their bodies, the moon set. By 10:40 P.M. Archie was shivering. They got up and walked, hiking in almost total darkness. Every now and then they bent over and touched the ground to make sure they were still following tire spoor, but by midnight they realized they'd been wrong. It was a zebra trail.

Again, Joel brought up conventional wisdom—when lost, stay put—especially relevant because it was night, and there was no chance of seeing a familiar landmark. Archie had never heard of such a ridiculous concept. It was too cold to sit still. They kept walking. Archie was adamant, and JB did not want to offend him.

At 1:30 A.M. they heard heavy footsteps. Without thinking, JB responded with his version of a rhino trill, a lone whoosh of expelled air followed by flapping lips. It was the basic rhino alarm call. Then, they just stood there, in the dark, at close range, waiting for a possible rhino to confirm its identity. Even Joel realized he'd just made a stupid move. They couldn't see the animal. Feet exploded, fortunately moving in the opposite direction and sounding characteristically like mountain zebras. Archie was terrified and convinced that Joel was crazy. Although Archie hadn't worked rhinos at night, when they were less likely to charge, JB had to agree that he'd been silly.

At 2 A.M. they found a rocky path with tiny ridges that could only have been made by tires. This time they were positive it was a road. They followed it. An hour later they surrendered to fatigue and stopped walking. Shivering, they pushed up against each other for warmth. Two kudus woke them at 4 A.M., and they sat huddled until first light. By 6 A.M. they realized where they were. They had walked in the wrong direction. By the time they reached the Land Rover they'd covered about 40 kilometers. I now knew why they'd arrived thirsty and exhausted.

16

Lions and Hyenas

[Carol]

Ngorongoro Crater, Tanzania: 1966: At 10:30 three sub-adult male lions were seen watching a rhinoceros with her calf. One lion got up and approached them. As he neared the animals broke into a run. The calf snuggled against its mother, who moved toward the approaching lion. The calf retreated and the lion pursued it, separating it from the female. The mother followed the pursuing lion at a steady trot, and the calf doubled back to the female. The adult immediately engaged the lion, who diverted his attention to her. The female whirled around with incredible speed and gored him twice in the centre of the ribs, using the anterior horn with quick stabbing thrusts. The lion rolled over, completely winded. The rhinoceros then gored the lion once in the centre of the neck, followed by another thrust through the base of the mandible, killing him instantly. The two lions had not moved during the entire proceedings.

That was John Goddard's only observation of a fatal interaction. Now, 30 years later, Ngorongoro has lost most of its rhinos. The cause: poaching, not predation.

Our sample of witnessed encounters between rhinos and lions and spotted hyenas grew to over 100. More than 95 were at night in Etosha. Unlike the lions in Goddard's tale, the few times that lions appeared to seriously threaten rhinos, they

worked together. The most intense encounter we saw occurred at our most populated study area, Calcrete Basin. Three lionesses and a thick-maned male pursued *Malawi* and her one-year-old calf from an open vlei to the forest's edge. The
rhino pair then reversed and raced back toward the open ground near the pool,
where the mother whirled back and forth and rushed in short dashes toward three
of the lions. The fourth, a lioness, moved toward the calf, who stood his ground,
head lowered. *Malawi* charged. With her tail erect, she powered her 50-centimeter
anterior horn toward the lioness's face but missed. Still battling, all six animals
disappeared into the dark woodland.

We wanted to follow but knew that headlights and the loud hum of our V-8
engine would frighten the rhinos and everything else in the silent vlei. So we
continued our work. Two nights passed with no sign of *Malawi*. On the third, she
and her calf drank. The lions had lost that one.

At Zebra Pan we saw more than 40 rhino interactions with spotted hyenas, all
at night. The largest number of hyenas we saw pursuing a rhino mother and calf
was seven. When the hyenas approached, *Susuwe*, the mother, just turned and
ran. Her calf followed. This time there was no active defense. The pair fled for
nearly a kilometer and vanished into thick bush. Six months later we saw *Susuwe*
again, still accompanied by her calf.

Other interactions involved different predators. Cheetahs were chased by female rhinos three times and once by a young male. In another case, a leopard
approached the benign old bull we called *Chama*. He lowered his head and stood,
as if waiting for the cat to impale itself on his spearlike horns. Seven full minutes
later, the leopard wandered away, disinterested in our visions of a spectacular
death.

A pattern of behavior began to emerge, one in which males and females had
different repertoires. Female rhinos were more aggressive toward spotted hyenas
and lions. Males were more likely to ignore carnivores. This greater female sensitivity persisted even when females were solitary. In situations where there were
no calves to protect, the fact that females were still more likely than males to
charge cast doubt on the idea that the purpose of their aggressiveness was to drive
possible predators away from calves. Some might argue that the behavior is innate.
In biological parlance, females were not currently investing in their offspring
through maternal defense. Perhaps they pursued a form of future investment:
driving away potentially dangerous carnivores so that they would be less likely to
attack in the future.

Two species of hyenas, spotted and brown, occur in Namibia, not closely related
to each other. The spotted hyenas are the largest, and they live in clans, which
enables them to hunt prey the size of zebras, gemsbok, and even eland. Brown
hyenas are lighter, weighing about 45 kilograms. They have pointier ears and long
dense brown hair, and their neck fur is lighter than the rest of their body. Despite
their size, brown hyenas do not actively prey on large ungulates. Instead their diet

consists of fruit, small vertebrates, and insects. On the Namib Coast, they feed on carrion, dead seals, fish, or birds that have washed ashore.

Aardwolves are also found in Namibia. They are diminutive by comparison, weighing less than 11 kilograms. Although thought to share common ancestors with true hyenas, they are in their own family. Their feet and muzzles are dark, a distinguishing erect mane adorns their neck and back, and their body fur color varies from tawny to yellowish. Among the most specialized of African carnivores, aardwolves feed almost exclusively on termites at night. Monogamous, mated pairs defend territories against interlopers.

Only rarely did we encounter this unique carnivore. We once saw a pair running on the gravel plains. At first Joel and I thought they were jackals. Archie wasn't certain. But their gait was different, and their shoulders were higher than their rumps.

Joel also saw them when out tracking in Damaraland. He and Archie were separated, and he was fording a dry river, when a noise in a nearby bush caught his attention. He was prepared for a rhino, but a dainty animal moved into view. Curious black eyes stared through a tangle of undergrowth. Ten seconds later a slender, striped body erupted from the bush, in flight. JB strolled along smugly, knowing he had seen something special. But the sensation didn't last.

Another extraordinary animal reared before him, too closely. White stripes coiled around a black body, it held its squat head 25 centimeters off the ground. Extremely poisonous, with tissue-destroying venom, a zebraslang (or a western barred spitting cobra), had zeroed in on him, poised only a few meters away. He reminded himself that the next time he sauntered through the silent desert in early morning light he wouldn't be so self-satisfied.

Large carnivores are big business in Africa, Asia, and elsewhere throughout the world. Although some people may be indifferent to their presence, more often than not dangerous predators elicit strong emotions. Tourists dole out huge sums to see cats. In Africa the major attractions are lions, cheetahs, and leopards. Hunters pay generously to stalk and shoot them, and local people appreciate their removal. Hereros, Damaras, and Owambos calmly point out life's inequities. White tourists, photographers, and conservationists call for protection when lions or hyenas are killed. But these are in areas outside wildlife preserves, communal lands where local peoples scrape out a living. On the white-owned farms, most predators have long been exterminated. Why, the communal people ask, should we not have the same privilege of safety and kill off the remaining predators when our stock or we are threatened? We had no answer.

Our most enlightening conversations with tourists came as we watched the animals with them at the Okaukuejo water hole. Most frequently we were asked where five species could be seen. The three most sought were elephants, rhinos, and lions; cheetahs and leopards were next.

Visitors' fascination with wildlife in general and predators in particular has not gone unnoticed. Tourism currently ranks among Namibia's top industries as an earner of foreign currency. The constitution states that "biological diversity . . . and living natural resources are [to be] utilized on a sustainable basis for the benefit of all Namibians, both present and future." Indeed, to be sure that a tourist-based economy continues, the government has placed a high value on understanding the ecological factors that regulate wildlife populations. In other words, they sanction research.

The logic is deceptively simple. If the animal numbers drop or species are not highly visible in Namibia, international tourists will be inclined to visit nearby parks in Botswana, Zimbabwe, or South Africa, where they will be assured of seeing diverse and spectacular wildlife. Unfortunately, at Etosha the ungulate populations have declined. A 1973 fence encloses the entire 22,700-square-kilometer park and prevents migration. Wildebeest and zebras have dropped from about 25,000 each in the 1950s to some 2,500 wildebeest and 6,000 zebras today. Since 1975, lions have been halved to less than 300, and cheetahs have become scarce.

It isn't clear why the numbers have plummeted. Limited food, disease, even the fencing are possibilities. So is predation. Perhaps lions and hyenas limit the plain's ungulates, which in Etosha include zebras, wildebeests, springbok, and gemsbok. An individual lion will, on average, eat about 4,500 kilograms of prey each year, a spotted hyena about 1,500. Some ministry biologists have suggested that if prey populations continue to drop, predators might have to be controlled to prevent them from eating themselves out of a food source. If they do, at some point their numbers too will drop. The economic implications are critical. Active control of carnivores would leave fewer predators for tourists to see, but not controlling them within a fenced reserve could leave less of everything.

It came as no surprise when Bill Gasaway and Andy Phillips spearheaded a massive project to study Etosha's carnivore and ungulate dynamics and to investigate establishing a wildlife-based economy for people living outside the park. The Namibian government welcomed the venture, and a joint agreement was signed with the U.S. Agency for International Development (USAID) and the Zoological Society of San Diego. Bill would run the field operation, and Malan Lindeque would direct actions from the Ecological Institute. Eugene Joubert would run interference from Windhoek. Andy, who had studied leguaans and cheetahs locally, and whose zoo-donated caravan we used in Okaukuejo, would coordinate activities from California. The million-dollar project had educational, training, and research components. Employment opportunities would be available for Namibian conservation students. Undoubtedly, the integration of science, conservation, education, and economic development was an ambitious attempt to further a noble cause: environmental sustainability in a newly independent and democratic country. What could be better?

* * *

We'd been in Africa nearly 16 months and soon would return to the States. But we were ambivalent. Joel had teaching and various committee responsibilities back at the University. I wondered how my mother's death had affected family dynamics and traditions. And the United States had traffic, pollution, and crime—some 20,000 murders each year. There was also the other side, access to education, technology, and our American culture.

In two months we'd be home. I showed a picture of our dogs, an elkhound and a blue heeler, to Sonja. She held it, stared at it for a few long seconds, and then pressed it tightly against her lips. Now almost three years old, she had somehow remembered them.

17

The Dead and the Brave

[Joel]

A black rhino lies bleeding, its body riddled with bullets. Minutes later the "whock, whock" of a panga breaks the morning silence. The distal end of the rhino's face opens exposing tissue and nasal bones. The horns are hacked away, along with a swath of meat from its back. The hungry poachers enjoy a meal. Yet the rhino still breathes. Flies swarm where blood flows. The female rhino musters enough energy to walk to the nearest water and collapses there. Her death is slow.

In 1847, Adulphe Delegorgue recounted his own experiences in *Travels in Southern Africa*: "We met up again with hunters . . . who had killed three rhinoceroses. . . . They told us wonderful stories of these hideous rhinoceroses, which they killed for the sole purpose of having them out of the way. . . . The Amazoulu Cafres have certain prejudices in regard to this animal which they passed on to us, so that the hyaenas and the vultures found themselves the only guests at the immense and delicious feast which took place every time we killed one."

Brutal scenes have been around for millennia. Recorded descriptions of the pleasure of killing are more recent.

Robert Struthers, in *Hunting Journal, 1852–1856, in the Zulu Kingdom and the Tsonga Regions*, described his sordid adventure: "enjoying the fun and nearly convulsed with laughter—The Rhinos were frightened. . . . and seeing the sportsmen . . . made for where we were. . . . I fired immediately & hit the other which tumbled over the body of the first—squealing like a pig."

George Eastman, operating out of Kenya in 1926, said indifferently, "The rhino's horns are nowhere as good as Audley's [another hunter], but I was afraid I might not have a chance at another, and I wanted the skin for my new library table top."

Hunting and killing can be very different events, even if the consequences are the same. In times past, massive slaughters were common, but in today's world hunters often give something back. Money sometimes goes to local people or management programs. In Africa as in other areas, legal hunting generates revenue and places an economic value on wildlife. Without this value, there would be little incentive to protect the animals, and they could be wiped out in a mere blink by growing communities of hungry people. Killing for sheer pleasure is another matter—squeezing a trigger, watching an unlucky target topple and die. The thrill is momentary and, to many, despicable.

* * *

For weeks blustery winds assaulted Etosha Pan and whipped soil from the alkaline surface. The soil billowed into an angry sky, mixing with residue from fields burning in the Kalahari and Caprivi. Atmospheric dust and overgrazing made the vibrant blue of summer a memory. On bad days the sun no longer set but vaporized, disappearing an hour or more before it should have. The dense air reminded me of childhood skies in Los Angeles. Winter in southern Africa continued.

On the murky horizon, the rising moon was invisible. More serious, our safety and data could be compromised. Reduced light made night obs precarious and individual recognition more difficult.

Just after dark *Somalia* arrived. He coursed the open calcrete in typical form. Ignoring us, he sniffed, then moved to new odoriferous ground for his next investigation. Elephant dung was bypassed, rhino dung was not. *Somalia* regularly plopped a few choice nuggets and propelled the juicy mass backward with adept rear legs. Swift pulses of urine jetted behind him. *Somalia* was alone, but it wouldn't be long until others joined him.

The curtain lifted on the night's rhino spectacular. *Malawi* and her calf *Malaise* entered Calcrete Basin's open vlei and walked directly to the water. Despite the veiled moon, our equipment amplified enough light to see. We recorded our observations from the Land Rover, Carol whispering as I wrote.

7:20 PM

Malawi	*Malaise*
Drink: 22 seconds	Head Up: 34 seconds
Head Up: 19 seconds	Drink: 38 seconds

Drink: 71 seconds, turn body from water   Head Up: 30 seconds, turn body from water

Two bull elephants interrupted my frenzied scribble. With their long legs they glided over bare soil toward the open water. Cunningham focused on them. I continued with the rhino cow and calf.

7:22   Somalia, 5 Rhino Body Lengths from the Elephants, Backsteps 4X.

Malawi	*Malaise*
Drink: 49 seconds	Head up, body away from water: 16 seconds
	Drink: 33 seconds

7:22   Separation Between Rhino Dyads and Elephants ca. 100 meters

Somalia	*elephants*
Head up and down 3 ×	NR (no reaction)
TU (tail up)	NR
7:23	EFlp—2X (ears flap, 1 animal only)
NR	
SO (stand only)	SO
7:24	
DOS (distance of separation) about 4 BL (body lengths)	
	Elephants move to water.

So went the evening's drama, actually a slow-motion lack of drama, as we plodded along gathering our behavioral information.

Key to our success was the night vision system, equipment that had operated wonderfully for months. The joints between the lenses and adaptors were weak. A 15-dollar part connected pieces now worth $13,000. To protect the soft threads I babied the system, placing it gently on pads when atop the Land Rover and packing it into foam-padded cases when we quit obs.

One late afternoon, as I repositioned the Land Rover for night work, 16 long, splotched necks poked above the forest canopy, watching. The sun shone weakly in the turbid sky as Carol gathered supplies from the tower. It was my turn to change the batteries in the image-intensifier unit, but the sound of "Daaaddy" broke my attention. I looked at Sonja and, without noticing, accidentally switched the unit on while pointing it at the setting sun. Minutes later I realized my mistake. I began to worry, remembering how a similar mishap back in Nevada had destroyed a colleague's unit. Cunningham knew from my expression that something was terribly wrong.

Helpless, we couldn't check the scope until the sky turned pitch black. While waiting for darkness, I reorganized spotlights and stared blankly at photos of rhino faces, memorizing the cuts and tears on the ears of individuals who might visit later. Finally it was dark enough for damage assessment. We inserted new batteries and were amazed to find that the image intensifier worked. The old batteries in the unit must have already been inert. Relieved, we finished our stint at the water hole.

At first light, after our fifth day of obs, we packed and left Calcrete Basin to head for Zebra Pan, which was almost 300 kilometers away; if we wanted to set up markers before dark, we needed to hurry. We regularly spent 5–6 days gathering data at each site. As we drove, we passed deciduous forests and drab green shrubbery where thin spiny acacias, called nebrownii, grew in impenetrable thickets. Except for noisy zebras and a few distant wildebeest, the veld was quiet. Dotting the road were moist balls of elephant dung.

Feces have been a valuable resource, an exploitable product, for hundreds of millions of years. Containing the undigested remains of both plants and animals, dung is treasured by organisms ranging from drab flies to brilliantly-colored butterflies. Even monkeys and horses will eat dung, a process called coprophagy. But the most spectacular fecal harvesters are dung beetles, also called scarabs; they include more than 200 species that feed on elephant dung.

In Kenya's Tsavo National Park, British zoologist Malcolm Coe estimated that elephants defecate up to 17 times a day. Before poachers decimated Tsavo's elephants, there were nearly two elephants for every square kilometer. Through simple calculations, Coe estimated that without the beetles every square kilometer of the park would be covered each year by more than 60,000 balls of dung! Just as astounding was the report of a single pile of elephant droppings in South Africa's Kruger National Park that contained 7,000 dung beetles.

With such an abundance of biodegradable material, the scarabs play an important role in community dynamics. When they break down the dung, nutrients are returned to the soil, absorbed by plants, turned back into green material, and reconsumed by herbivores. It's just one of many complex life cycles in an ecosystem with large mammals. In the very dry areas of the western Namib Desert, dung beetles are less common, and feces can be broken down by termites.

*   *   *

Among the many mammals that delight in eating fresh plants are humans. Carol and I craved greens, unusual in markets away from Windhoek. During our first week in the field in 1991, the husband-wife photographic team Des and Jen Bartlett had explained to us that they solved the problem by raising sprouts even when traveling. As students, and well before we knew each other, Carol and I had done the same, she in Santa Barbara, California, and I in Boulder, Colorado. In Africa we soon realized that, like the Bartletts, we needed a mobile market.

As always, we were constrained by our water supply, unwilling to waste it even for a good cause like converting lentils to sprouts. But after months without a green salad, we desperately craved variety. Crackers, sardines, and even soup could be much improved by adding sprouts.

Our dilemma was finding a spot for jars in the overburdened vehicle, a problem we never satisfactorily solved. Jars tipped, water spilled, and sprouts desiccated in the intense sun. I thought of my mother, especially her disapproval of my untidy room. If she could see the disorderly Land Rover, it would confirm her suspicions that some people never change.

One day, while making fresh bread, Carol chimed, "Let's make a sprout and avocado sandwich. This avocado hasn't been bruised too badly."

Great idea! Like a dog salivating in anticipation, already I could taste the warm bread with komkommers (cucumbers), tamaties (tomatoes), and sprouts. The bread cooled. I went to get the sprouts. Twenty minutes passed as I searched the Land Rover. Somehow, somewhere, the sprouts had vanished. Frustrated, we ate the sandwiches sproutless. For the next week we wondered about the missing jars. Even after we left Damaraland and downloaded the Land Rover, they never reappeared.

Six days later, I shifted a 20-liter water jug from behind the passenger seat. The top was loose, and the water level was low. Water must have leaked as we drove. On the soggy rug below, sprouts grew everywhere. Although we had a new supply of fresh greens, I could see my mother's look of disgust and hear her say, "How will you ever find anything if you don't keep your room clean?"

We never found the jars.

When our work began, there had been only a handful of observations of interactions between rhinos and potential predators. Little was known about the behavioral responses of rhinos to hyenas, lions, and humans, or whether horn size was important in deterring predators. Archie had told us that poachers do not prefer males or females because it was so difficult to find rhinos at all. Anyone bent on killing would not waste their efforts by forgoing a shot.

For most mammals, the situation is radically different. Elephant males are larger than females, and their tusks are larger. During the 1970s and 1980s poachers consistently chose males; their ivory was worth more than the smaller amount carried by females. As a result, fewer adult males survived, and the sex ratio became heavily skewed.

Such imbalance is the typical mammalian pattern. From diminutive ground squirrels to colossal whales, and from Asia to Africa, Australia to North America, females outnumber males in many species. Kudus, lions, and baboons all fit this scenario; so do polar bears, kangaroos, and moose. Biologists are keenly interested in this pattern, because it persists even in the absence of human hunting.

In 1871 British naturalist Charles Darwin provided insights about the widespread skew in adult sex ratio and, at the same time, offered suggestions about the mysterious relationship between sex differences in body size and weapons: "When the males are provided with weapons which in females are absent, there can be no doubt that they serve for fighting with other males. . . . They must also be often exposed to various dangers, while wandering about in eager search for the females. . . . the lion in South Africa sometimes lives with a single female, but generally with more, and, in one case, was found with as many as five females."

Darwin went so far as to speculate how competition among males for females resulted in differences in breeding sex ratios: "The practice of polygamy leads to the same results as would follow from an actual inequality in the number of the sexes; for if each male secures two or more females, many males cannot pair."

In reality, for species in which males carry horns or antlers or are larger than females, there is a cost: early mortality. The cause of this higher mortality is not totally clear. Males usually cover larger areas, wandering more than females, and they tend to be less vigilant toward predators and fight more than females. Males also die at younger ages, for any number of reasons—injuries, disease, starvation, and outright predation. Even male humans tend to live shorter lives than females.

For black rhinos the entire scenario differs. Unlike elephants, and other species in which males are larger than females, rhino males and females are about the same size. This we knew from our measures of immobilized animals and from results reported by Canadian and South African researchers. Furthermore, based on rhino skulls and horns that we had tracked down in vaults, in Khorixas, and in western Etosha, we now knew that males and females had horns of similar size.

If Darwin was correct, and larger body or horn size equates to higher mortality among adult males, we speculated that the sex ratios of black rhinos should be equal. Our data were nearing the point where we could make a first stab at evaluating this idea. The prediction itself was simple enough. In the absence of human predation, males should be as abundant as females. We used data from our two major study areas, the Kaokoveld and Etosha, to test it.

At both sites we found that adult males outnumbered females, but the differences were small. Statistically, there were no differences. Nevertheless, our limited information suggested that, unlike most horned species, in at least two Namibian populations the sexes were equally abundant.

To supplement our data sources, we turned our attention elsewhere, using information on adult sex ratios from populations in Zimbabwe, South Africa, Kenya, Tanzania, and Zambia. In most, but not all, of the populations, males were slightly more abundant than females. Where females outnumbered males, poaching had been profound. Males outnumbered females in 90 percent of the protected populations, whereas males were more abundant in only 20 percent of the populations with intense poaching.

Three possible factors might account for these differences in adult mortality. First, as with elephants, poachers may prefer males. Although Archie didn't like this idea, because males don't have larger horns, it could be that poachers simply encounter males more often than females rather than showing a preference for them. We examined this possibility by checking to see which sex we encountered more. There was no difference. A second explanation is the existence of a sex bias at birth. This was a tricky proposition to evaluate. Few data are available on birth sex ratios in wild rhinos. So we checked birth sex ratios spanning a 17-year period using the International Studbook for African rhinos, and did similarly for a 39-year time frame with information from the American Association of Zoological Parks and Aquaria. We assumed birth sex ratios in zoos would be similar to those in the wild. And, indeed, captive animals give birth to slightly, but not significantly, more males than females. A third reason for sex differences in adult mortality centered on behavior. If females were more flighty than males, then males would be more vulnerable to poachers. This hypothesis we could assess.

We had recorded how often and how far males and females ran after photographing them at close range. Our approaches and methods of gathering data were the same for each sex, so there shouldn't have been any biases in one direction or the other. We also made a critical assumption, that our method of tracking rhinos was no different than that used by poachers. We couldn't test this assumption, but both Archie and the Owambo trackers we worked with in Etosha told us that they all operated the same as poachers—tracking by foot until at close range.

Our findings were striking. Females ran farther than males—a sex difference that persisted irrespective of habitat. In Damaraland's dry rivers, females ran on average four kilometers, compared to one kilometer by males, before resting. On gravel plains, our disturbance caused the animals to move even further, most probably because of the lack of cover. There, females moved more than six kilometers before resting, males less than three and a half. Such flight distances in the absence of pursuit might be the largest for any solitary mammal. However, we also found males that moved less than 400 meters. At times *Mad Max* and *Pinocchio* moved less than 100 meters.

Now that it was clear that male and female rhinos reacted to us differently, we also wanted to know whether the sexes varied in their responses to dangerous predators like lions and spotted hyenas. We didn't see any interactions in the Namib Desert, but in Etosha our sample built to 144 encounters. The data were illuminating. Adult females were more vigilant or more aggressive than males in response to both those potential predators. It didn't matter whether females were alone or with calves; they attacked lions and hyenas more than males did.

The interactions with us helped to explain why a sex ratio shift occurs in populations with human predation. Adult males are relatively stolid and females more sensitive. By not running far, males are easier prey for poachers. They respond to humans much as they do to lions and hyenas. Females, on the other hand, were reliably excitable. They either fled or charged, behavior that substantiated our greater fear of being impaled by females. In fact, I unwisely felt myself relaxing my guard if we were working males.

Why the sexes behave differently was another question. Females might be expected to show greater variation because they invest more heavily in offspring than males do. But our data concerned only solitary females and solitary males. In this way, calves could not confound the analyses. This is why we wondered whether solitary females respond more to dangerous predators as a form of future investment: consistently harassing lions or hyenas to make them less likely to attack in the future. Unfortunately, without data on individual lions and hyenas, these ideas could not be tested.

With our rhino data analyzed, and after much discussion with the small cadre of Etosha biologists, we drafted a manuscript and circulated it to Malan Lindeque, Eugene Joubert, and Bill Gasaway, planning to show it to colleagues in North America when we returned there. Much of rhino biology left us puzzled and with many unanswered questions. We knew that, like other polygynous mammals in which males compete for access to females, male rhinos died in combat. We wondered how adult sex ratios remained balanced unless other mortality factors were operating on females. What these might be eluded us. We also wanted to know which males sired offspring, knowledge that would help us understand some of the implications of horn removal. And, although our sample sizes on interactions between rhinos and potential predators were large relative to those in past studies, an even larger sample would improve statistical analyses.

18

Concrete Corridors

[Carol]

Having finally begun to understand something about rhinos, we debated the merits and pitfalls of making management suggestions. Actually Joel mulled it over. I was his sounding board. We now were sure there was sentiment against us and our project. One ministry officer had accused us of shooting guns indiscriminately, even though the only shots ever fired had been when I was charged and treed by a rhino the year before. JB had never even carried a gun in the desert. Still, the accusation echoed through the ministry. Rumors grew. Because we weren't seen often, people claimed we rarely worked, or that we sat in chic cafes in Swakopmund on the coast. We couldn't answer the gossip, short of sending out a weekly newsletter, so we tried to ignore it, focusing instead on logistics and rhino biology. We had thought that although dehorning was controversial, Malan had welcomed our proposal, and we had support. But now we inched toward thorny political ground.

To evaluate dehorning, we needed to know about the status of populations before dehorning. This meant, beyond finding records, we had to have confidence in the methods used to gather them. Through our work in the vaults, biologist Peter Erb's records, and information provided by Garth and Archie, we discovered that three potentially breeding females south of the Veterinary Cordon Fence had been transferred to reserves elsewhere. Three, by itself, isn't a large number, but it was 25 percent of the female population.

We were squeezed uncomfortably between choices. It seemed unwise to question prior decisions, even when those might jeopardize the viability of Namibia's dehorned rhino population. Only 9 potentially breeding females remained in the area. But, the animals had already been removed. If we continued to work independently and quietly, we couldn't be considered agitators. However, by saying nothing, we'd be guilty of a de facto sanction of questionable policies shaped by shoddy data. If, on the other hand, we came forward, we'd enter uninvited into a controversial arena.

Troubled, we sought advice from trusted colleagues within and outside the ministry, but ultimately we had to decide what we could live with. If we wanted to be unobtrusive, it might be best to lie low and be silent. Other than Eugene Joubert and Malan, no one seemed interested in our findings. And, unfortunately, Eugene told us he wanted to leave the ministry and that he might soon be gone. Time passed, and we said nothing.

Then, one day when we were back in Okaukuejo, Malan asked Joel for his thoughts about rhino management. This was the opportunity. Joel drafted a memo.

April 23, 1992

Malan—a few thoughts on rhino viability in the Kaokoveld
1) Accurate Long-Term Data—A Necessity
   . . . individuals must be reliably identified. This cannot realistically be achieved given the current census techniques. . . . Given the variation in quality of past and existing photos, serious problems lie ahead, jeopardizing any credibility in estimates.
2) Population Viability Analysis Sorely Needed for Rhinos South of the Red Line [Veterinary Cordon Fence].
   — Many of the animals in the Doros Crater area are derived from one female.
   — Attrition of females in the Poacher's Camp Area . . . makes one wonder how long they can persist
   — If dispersing subadults are to be rescued regularly, then perhaps adequate space no longer is available. If the goal is to use the young animals south of the Red Line to repopulate other areas, what happens once the few remaining breeding females reach senescence?
Obviously biology cannot be divorced from the real world. But there must be options. . . . A plan of action . . . needs to be worked out . . . which will result in higher assurances of the long-term persistence of Damaraland rhinos. . . . For whatever this is worth, I do not at all mind if you share this with others.
Best—Joel

Nearly two more months passed. The ministry, with Save the Rhino Trust, continued their survey of the entire Kaokoveld, identifying every animal. The area was more than four times as large as the subregion that we worked.

Pierre du Preez, a large, jovial, bearded South African biologist, was in charge. As rhino coordinator for the ministry, he led teams into and out of the Kaokoveld during the grueling census. His drinking was legendary, his competence unquestionable, and his knowledge of rhinos unfailing. Pierre would be happy to talk with us. We anxiously awaited the results of the census to find out not only how many total rhinos existed but how their identities stacked up—particularly whether Pierre would cover up past errors or correct them. Sooner or later, we'd see Pierre and ask.

* * *

Back at Calcrete Basin, the stench couldn't be filtered from the aroma of early morning coffee. Wafted on the dusty breeze was the fetid odor of yet another dead elephant, the fourth we'd seen in a week. Concentrating on our cereal, we attempted mind control; "ignore" became our mantra.

At lunch we dragged together two cracked plastic baby chairs and the remaining camp chair for our big meal. Soni wasn't happy. With each wind shift, smoke blew in her eyes. She finally jumped up from her chair to climb into the Land Rover. But as she turned she shouted, "Uh oh." We turned to see a huge bull elephant, 30 meters away, eying us on his way to drink. "Uh oh" was right. Her parents were totally unobservant.

I moved quickly to Sonja and lifted her. She whimpered. Her eyes trickled the light flow of tears before the flood. Both JB and I told her, "It's okay. He's a nice elephant, he just wants to drink."

Luckily, it was true. He drank, sprayed himself, wandered about, returned, and stood next to the water. He was a welcome contrast to the high-strung elephants we sometimes encountered.

During the night a single lion roared. Nineteen elephants visited and left, leaving the pool with 2,400 fewer liters (about 600 gallons) of water. The rhinos that evening were confusing. Of the 15 individuals that appeared, we knew only 11. Frustrated, we went to bed early, at 2 A.M. It was less than 40°F, and with the wind chill the temperature felt below freezing. In the morning we drove to Zebra Pan to continue our night watches there.

The cold winds continued. One night we drifted to sleep with thoughts of an arctic landscape, frosty and shimmering below a full moon.

At 7 A.M., "Maaama, Maaama, come down" roused us.

"What day is this?" said Joel, climbing down to give me a few more restful minutes.

"I don't know. You mean the date? Last night was the full moon." *Leave me alone, I want to sleep.*

"Then why is there a half moon setting?"

"You're nuts, Berger. Over the edge." I was too sleepy to even open my eyes.

"Oh darling, what do you mean?" His voice was lively, daring me to look. The moon, only half full, was disappearing in the western sky.

Namibian skies were getting the best of us. How could this be? We'd gone to bed under a full moon. Were we back in the twilight zone of southern Africa? It brought back our first day in the field, befuddled by the space shuttle in the western sky.

Sleepily, I made up a story. "Joel, the only rational explanation is that we slept for a week, that's why Soni yelled, she's starving."

"Sure, Cunningham, sure; the moon wouldn't be setting at sunrise if you were right."

We enjoyed breakfast and mused about the eclipse.

Back at Okaukuejo, the rhino *Kipsigi* charged lions four times during a two-hour period. She chased the aggressive old bull, *Kaoko,* but melted into the trees when 40 elephants showed up to drink. Tourists laughed, thoroughly enjoying the nightly entertainment. People from France, Germany, Canada, South Africa, and Kenya peered through the night vision equipment and asked about our work. These tourists loved Africa's wildlife parks and watched with awe as the animals stirred their imaginations. Music blared from the "location," where the Okaukuejo workers, mostly laborers and their families, lived. Few tourists noticed.

Our 1992 fieldwork came to a close.

* * *

Africa was frayed with unrest. Liberia, Angola, and Sudan were out of control; so were Ethiopia, Somalia, Mozambique, and South Africa. Impending violence threatened elections in Kenya and Zambia. America beckoned.

We had missed events. Riots in Los Angeles had left 58 dead. Gorbachev was out, and the Soviet Union had collapsed. Yugoslavia was in civil war. Indira Gandhi had died in a bombing. The Clarence Thomas and Anita Hill proceedings had mesmerized the American public. George Bush would run again for the presidency. This time the challenger would be an unknown named Bill Clinton.

We thought of home—snowy woods high above the arid desert, an aspen grove hidden on a secluded mountain—memories from former field sites, and our early relationship. The thought of reentry into "civilized" America was overwhelming. Coping with the frenetic pace, concrete corridors, and the halls of academia was not something either of us wanted to think about, at least not now. The cool mountains and hot deserts of Nevada would be our refuge.

* * *

At 10 A.M. Pierre du Preez entered the Institute office where Joel was working. He wanted to discuss the results from the Kaokoveld rhino survey.

"Hey Pierre," called Joel, "how accurate do you think the census was?"

"I don't know, Joel—let's compare our information and find out."

I was preparing tissues from the ear of a dead rhino for later DNA analysis; I joined them later. Photos, handwritten notes, and maps covered the desk and part of the floor. They had been at it for three hours. Joel announced that he and Pierre were in agreement on about 90 percent of the resightings. Pierre corrected him, smiling: "Ninety-five percent."

Not only was it important that JB and Pierre agreed, but they had finally resolved the identities of many of the mysterious desert rhinos. They exchanged information on dates and locations of rhino sightings. By comparing their photographs of ears and notches, it was now clear that animals dehorned more than three years earlier could not have been identified correctly. Ear notches had either been noted in the wrong spot or not noted at all. Tragically, 50 percent of the animals dehorned in 1989 had been incorrectly identified in Save the Rhino Trust and ministry files. These animals could not have been monitored accurately.

Pierre now understood why Joel and I were so interested in having confirmed identities. Finally it would be possible to check on the performances of known individuals before and after dehorning. To help us, Pierre also provided the estimated birth dates of new calves born to two dehorned mothers.

Then, self-deprecatingly as if to indicate that he was on top of the records, Pierre stood, saying, "What does a fat Afrikaner know?"

"Plenty," smiled Joel.

With that, Pierre extended his hand, and said, "Have a good time in America."

* * *

Tires screeched as we landed at John F. Kennedy Airport in New York. From Windhoek to Johannesburg, to the Cape Verde Islands, and then from New York to Los Angeles, Sonja had been a champ, talking, looking at books, and playing with her doodle drawing set.

The subsequent conversations about Africa floored us. Even educated Americans seemed to know little about Africa in general, let alone rhinos or the horn trade

An acquaintance asked, "Did you bring back anything good?"

"Like what?" We thought they meant baskets or artwork.

"Rhino horn. I hear it's worth a good deal of money, like elephant ivory."

Shocked to hear that both horns and ivory were illegal because the animals were endangered, they countered, "Well, why don't they use tranquilizers like on the TV shows?" Indeed, that's what southern Africans had in mind. We were about to explain the complications, but the splash of a body diving into a swimming pool cut the conversation short.

The outlook from the States seemed to be that most poachers were high tech, roaring through the bush in monstrous jeeps fully equipped with roll bars and spotlights. "Just shoot the bastards" was a common phrase. The lives of rural Africans were incomprehensible. How could people live on dirt floors and have no refrigeration? Or how, in some of the cities, could unemployment reach 70, 80 percent?

Home again in northern Nevada, things were no better, no worse than before. Our dogs were healthy. We settled in. Sonja, almost three years old, was in serious need of socialization. We, and the Land Rover, had been her security, constants in a gypsy lifestyle. Although she knew about toilets, we had to make sure that she wouldn't just drop her pants. As for her constantly unkempt hair, Berger and I were teased about "the feral child." Preschool would definitely be in her immediate future.

A handful of colleagues were interested in our work. Most asked about our "vacation."

Autumn turned to winter. We analyzed data. The snow fell; with each storm, the layers grew higher between the pines. We shoveled and walked between walls of snow, remembering walls of brush in the distant veld. January would soon arrive. Then we'd be back in Africa.

19

Of Science and Ecology

[Joel]

Science," my notes read, "is the acquisition of truths that bear on natural phenomena. At universities and colleges scientific study should result in new knowledge, some useful, some not."

"Borrr-ring" is what many people say of science—"okay for anyone but me." As a young student I had felt the same, until the congestion and unending chaos of Los Angeles closed in, and I felt a need to escape. Although summer beaches were spectacular in those days, the—sixties—the people multiplied, and the winter surf was cold. Unpopulated places were seductive. I fled to the desert where, gradually, my interest in the biological world developed. Some sights disgusted me—tortoises shot while crossing roads, coyotes and other species poisoned. I wanted to make a difference. If the study of science could get me to areas where plants and animals had not been erased, then occasional hardships were worth enduring. At Northridge's California State University my undergraduate major changed from sports to biology. I was 19 years old.

As a discipline, science has something in common with journalism; both require objectivity and facts. The similarities end there. Good science requires hypotheses to be tested and falsified, adequate sample sizes, and repeatable events. Experiments carry the burden of proof. The process of doing science has not changed.

What has is that more people are interested in how science is applied to environmental health and conservation. Reports in *Newsweek, Time,* and specialized magazines, as well as television and the cinema, have fueled an appetite for information. Some material is biased, and because few people know what to believe, credibility is an issue.

Universities and colleges have educational programs that emphasize science in the public interest, making the process of scientific discovery coherent and evaluating information possible.

Indeed, how the public can continue to perceive science as dry is puzzling. Science involves "politics, sex, wine, movies, teamwork, rivalry, genius, stupidity, and virtually everything else that makes life in the lab and out something less than perfect and a great deal more than dull." So declared J. D. Watson in his 1968 book *The Double Helix,* which details the discovery of DNA. Like life, science is diverse.

Scientific discoveries such as infant-killing in species as varied as lions, blue gill sunfish, and poison arrow frogs came about not because biologists had initially looked for the phenomenon but because of chance observations. Then the scientific method kicked in. New questions were raised, and investigations of the causes or consequences of behavior began in earnest. Studies were designed to test predictions stemming from adaptive and nonadaptive hypotheses. Laboratory and field experiments followed. Unadorned, natural history was the springboard, key to developing projects in biological conservation.

* * *

Our third field season was about to began. New equipment was ordered—lenses, upgraded night vision gear, solar panels, and a laptop computer. Our study design was well in place. Each of the three Namib Desert sites served as broad-based experimental areas. All were ecologically similar, receiving on average less than 75 millimeters of rain annually. Springbok River had dehorned rhinos with spotted hyenas, Doros Crater had dehorned rhinos with no dangerous predators, and at our third site, horned rhinos lived with hyenas and infrequent lions. Etosha, although different ecologically, allowed us to investigate the consequences of natural variation in horn size on different patterns, including social behavior and reproduction where predators were abundant. The Zebra Pan, Okaukuejo, and Calcrete Basin regional populations could be compared with one another. We had toiled to find sites where comparisons were feasible. Horn removal offered an experiment.

There were drawbacks. Field studies rarely serve as perfect experiments. Animals vary in age, in their history, and in their genetic makeup. Although we used populations north and south of the veterinary cordon fence, slight, almost imperceptible, ecological differences will always exist among study areas. And not being able to watch every animal every day limited the amount of data that could be gathered.

In a perfect world, there would be no endangered species and no need to prioritize the study of one organism over another. Imperfect knowledge wouldn't exist, nor would populations be so small that only a few individuals provide information. But as every human soon comes to realize, including scientists studying endangered species, the world is flawed.

As refuges for species and ecosystems, national parks also serve as bastions of knowledge. But scientists today question whether parks stave off extinctions. Many protected areas are simply too small to guarantee persistence. With these ideas in mind, Carol and I wrote a paper on the viability of mammals in the Namibian national parks. It pointed out a need for more space if wildlife conservation was to succeed. We wrote the paper because Malan was organizing a symposium and asked us to contribute. Too many times foreign scientists have published their findings in journals not readily available to host countries. We saw this as a chance to communicate with local scientists and naturalists about recent principles in conservation. Malan made valuable suggestions, and we invited him to be a coauthor. Then we submitted the manuscript to Namibia's own scientific journal, *Madoqua*.

In it was a section on how long small pockets of black rhinos might persist. We knew that scattered groups of six or fewer rhinos occupied disjunct desert regions, parts of the Twyfelfontein, Brandberg, and Skeleton Coast reserves. Some rhinos undoubtedly used these areas seasonally, but none of the areas on their own appeared large enough to sustain viable breeding populations. Collectively, the reserves in conjunction with communal lands offer an area where all animals may be treated as one large population, assuming that movements are not blocked. However, because all of these areas may be affected by mining operations, transient human populations, and subsistence farming, the potential for the fragmentation of rhino populations remains.

We deliberately avoided making specific predictions about how long we expected rhinos to persist, because of the great uncertainty of human involvement. However, we closed the paper by noting that, even without the threat of poaching and other human disturbances, any serious attempt at conservation must incorporate the effects of humans rather than simple biological models alone.

A different paper with the bland and typically scientific title, " 'Costs' and Short-Term Survivorship of Hornless Black Rhinos," was likely to get us in hot water. This one we submitted to the America-based international journal *Conservation Biology,* which has a readership in more than 40 countries. We, in this case, included Archie and Malan in addition to the two of us. Having Archie as a coauthor made perfect sense; without his brilliant tracking we would never have found so many rhinos. Malan was quite a different matter. An official collaborator, he had arranged access to the vaults so that we could measure horns, and his suggestions had improved the paper.

But having Malan's name on the paper represented something bigger. Although Malan and Namibia's rhino and elephant programs were known within southern Africa, on the outside little was known about them. If the international conservation community was to recognize legitimate advances being made in the management of these controversial flagship species, perhaps the collaborative efforts between foreign and Namibian scientists should be articulated more clearly. What better way, we thought, to acknowledge Malan's substantial contribution? Admitting that horns were growing faster perhaps than expected, and that poachers may not be choosing rhinos with larger horns, would enhance the credibility of Namibia's rhino program. Of course, the final decision was Malan's. He agreed to be a coauthor.

An earlier version of the paper had been rejected by the British journal *Nature*. They had, however, been willing to take a 500-word piece—a simple summary of the essence of horn regrowth—as a scientific correspondence. It appeared in January 1993 and offered a rough idea about horn size and mass in dehorned rhinos. On average, front and rear horns had regrown six and three centimeters per year. Because we also had weight and size measurements of horns in the vaults and could estimate horn diameter and length from our photos, we also calculated the mass of regrown horns.

With the weight of front and rear horns combined, dehorned animals were putting on about half a kilogram per year. Three years after dehorning, the regrown mass weighed one and a half kilograms, or just over three pounds. On the Asian black market, the price per kilogram varied from $3,737 in Taiwan to $16,240 in Hong Kong. Obviously the regrown rhino horn was valuable, but we didn't know if Asian values would also apply to regrown horn or how much money the actual poachers received. What seemed to matter was that horns regrow rapidly.

Besides estimating the value of regrown horn, it was equally important to know whether poachers preferred rhinos with larger horns. If they did, then smaller-horned rhinos, and certainly dehorned ones, might be spared simply because their horns would be worth less money. We had no direct data on the issue, so we made assumptions. First, we presumed that horns confiscated from poachers were representative of the horns from living populations. There was no reason to expect that the samples in the vaults were in any way unusual. Second, we believed that the horns in the vaults were labeled accurately—that those designated as specimens from police had been shot and had not died naturally. Third, we needed to know something about the distribution of horn sizes in live rhino populations. We assumed that poached horns came from populations whose horn size profiles did not deviate widely from populations without recent poaching. If differences existed between poached and protected populations, we could not safely deduce anything about choices made by poachers.

Fortunately, comparisons were possible. We had data on the distribution of horn sizes in all the populations we studied, either because we had estimated horn size from our photographs or, in the case of dehorned populations, we had measured horns in the vaults. Our analyses revealed no differences between the poached and protected samples. Poachers didn't distinguish between rhinos with big and small horns. Archie was right. His feeling that poachers shot whatever they encountered was borne out by the data. But the practical implications of this conclusion did not bode well for dehorning as a conservation tactic.

Previously Malan had claimed, in a 1990 article in the *South African Journal of Science,* that "the objective of the dehorning programme is to deter poaching in an area where it had flared up and could not be controlled by other means." Our findings on rapid horn regrowth and lack of poacher choice now suggested that even animals with small horns might be vulnerable to poachers, and therefore should be dehorned regularly if they were not to be at risk of being poached. A similar conclusion had been predicted through a model by three British biologists, E. J. Milner-Gulland, J. Beddington, and N. Leader-Williams, in a paper that predated ours.

Within Namibia, and prior to the piece in *Nature* some were quick to point out that no dehorned rhinos had been killed, thereby implying that the dehorning tactic was in fact working well. The media loved it. Although rhinos were already in the international spotlight, given the controversy about dehorning and the vast amount of prior speculation, stories abounded. Now more international dollars could be garnered for hands on action programs like dehorning programs.

Left unsaid, of course, was that horned black rhinos in Namibia had not been killed either. Poaching was at a lull, or it had stopped altogether. There were multiple explanations for why poaching was down, but none could be tested in a scientific way. Speculation raged. But to us it seemed wrong to allow the world to think that dehorning had crushed epidemic poaching. Although we had written reports to the Namibian government and met repeatedly with officials, we also felt that as researchers we had a responsibility to report our findings. The piece in *Nature* would at least indicate that there was more to the story.

Many had hoped that dehorning might offer rhinos a future, since conventional protective strategies in other countries had clearly failed. Our concerns with respect to dehorning and poachers were unacceptable to those in the conservation communities of Zimbabwe and Namibia. They wholeheartedly accepted the tactic and promoted the possibility of developing a legalized horn trade.

The demand for rhino horn continued and, like South Africa, both Namibia and Zimbabwe had stockpiled horns in vaults. Legal trade could be lucrative. Not only would the horns already be in hand, but, theoretically, a free market with horn sold openly would bring down the inflated black market prices. That horns had value outside of Asia was not lost on the southern African countries. Continued rhino dehorning, regardless of poaching pressure, might mean money. Furthermore, an application on behalf of an American to legally import horns taken from a Zimbabwean rhino during dehorning was pending with the United States Fish and Wildlife Service (USFWS). If "darting safaris" could be encouraged legally and the removed horns brought back to overseas markets for show, then host countries could make money while simultaneously reducing the incentive for poachers to shoot the hornless animals.

In late 1992 the USFWS solicited formal comments on conservation actions for rhinos. The New York Zoological Society's Wildlife Conservation International, the World Wildlife Fund, and a host of other NGOs responded. Additional forums for the discussion of conservation methods would also exist. A Standing Committee for the Convention on the International Trade of Endangered Species (CITES) would meet in Washington, D.C., in early March 1993, and a June conference was to be held in Nairobi, where possible donors could commit funds. Dehorning was one little piece in an uncertain puzzle, the completion of which might halt the extinction of Africa's rhinos. To solve our section, it was critical to learn more about the biological effects of dehorning. Both Janet Rachlow's project on white rhinos in Zimbabwe and ours on blacks, in Namibia, took on renewed vigor.

Year of the Scorpion
[1993]

20

The Europa Hof

[Joel]

January. Pacific storms pounded the mountains that separate Nevada from California. Outside our home up in the Sierra Nevada Range heavy snow covered the sagebrush and manzanita, and the pathway from our front door was lined by snowy walls more than two meters high. With our Volkswagen van parked in the paved cul-de-sac, we trudged beyond the icy driveway and the unplowed dirt road, tugging trunks destined for Africa. Namibia's warmth and sand would be a shocking contrast.

Old and new equipment piled against the wall in my parents' Los Angeles home. The third generation night vision scope, "Dark Vader, Series 4501 . . . Vision Goggles" wore an ominous label that we'd just discovered. Apparently, the 7,300-dollar item needed clearance from the United States government to leave the country. Nine days remained before our departure.

Unclassified defense articles were taken seriously by the Bureau of Politico-Military Affairs at the State Department in Washington, D.C. In a panic, I called to request an application and, after mailing it back, inquired every few days about final clearance. No answers came. We finally left, equipment in hand.

I knew the State Department would not be happy, even if I rationalized the action by saying that there had been no problems taking a Litton night vision system to Africa during previous trips. But how could they deny something that was advertised and purchased legally in the United States? When I explained via telephone from Los Angeles that I had brought similar equipment into Namibia two years ago, Carolyn Love, the chief bureau officer, in Washington, D.C., briskly told me that it had been done illegally.

"Why wasn't I notified of a problem?"

"Because you didn't ask."

"I filled out equipment lists and shipped through the Smithsonian. Wasn't that legal?"

"You didn't apply through this office."

"Oh." Clearly there was no point in arguing with her.

* * *

Billowy white clouds trailed each other above the expansive Kalahari Desert. Below, fossil riverbeds and waterless pans stretched to the horizon. Jagged ridges clothed in summer's green signaled our entry into Namibia. As we stepped out of the plane, hot air surged into our faces.

"Welcome to Namibia," said the customs official. "Business or vacation?"

"Business." I fumbled for paperwork. "Here is our stamped letter from the Ministry of Wildlife, Conservation, and Tourism."

A rubber seal slammed onto our passports, the sound echoing through the hollow room. Curious officials in blue-and-white uniforms assailed us with a barrage of questions about our six trunks. Then we were free—until I noticed the length of my allowable stay: "21 days." For mysterious reasons, Carol's passport carried a 90-day stamp. We shrugged off the inconsistency, knowing that the needed extensions could only be applied for at the Ministry of Home Affairs.

It had taken an exhausting two and a half days to reach Windhoek, enduring crowded terminals and planes with a wriggling three-and-a-half-year-old, nine movies, and nearly a dozen premade meals. Halfway through the trip, Carol's face turned red and erupted in hives. We'd forgotten about monosodium glutamate in prepared foods. At least her allergic reaction didn't get worse, and she asked for the less dangerous vegetarian or low-sodium meals for the rest of the journey.

In Windhoek at last, friends updated us on recent events.

We needed to find out about Archie. Rumors had him back in jail, and Sharon Montgomery, the director of education of Save the Rhino Trust, had written to us that she was unsure of his whereabouts. We were concerned. If Archie was in jail, we'd have to search for desert rhinos without him. There was other unsettling news. Eugene Joubert, one of our staunchest supporters, had left the government, taking a job in Riyadh, Saudi Arabia. Others had also opted for jobs in the more lucrative private sector, including people who had helped us in the past—Etosha veterinarian Peter Morkel, the chief of management at Etosha, personnel in Etosha's Anti-Poaching Unit, a permit officer in Windhoek, and Simon Mays in the Caprivi Strip.

We reported first to the director of the ministry, Polla Swart, and Joubert's temporary replacement, Henti Schraeder, briefing them on our research plans for the upcoming year. As in the prior two years, our research permits were immediately granted. We could proceed to Home Affairs for additional clearance. Polla and Henti also thanked us for copies of the *Nature* paper on rhino dehorning, saying that it would be useful in developing plans for future rhino conservation.

The head of management, Dannie Grobler, was less effusive. It wasn't the paper that bothered him, since he had questioned the usefulness of dehorning four years earlier. What concerned him was that informants from northern Namibia had divulged that people in Angola's southern war zone had sought information about the ministry's personnel changes. The sentiment was that organized poaching might begin soon, and already indications were bad. A black rhino had been shot on the Angolan side of the Caprivi, and three Namibians and a Lebanese man had been arrested in Johannesburg with rhino horn. The poaching of white rhinos at Waterberg Plateau continued.

But there was good news. In Etosha, Bill Gasaway's million-dollar project on predators and prey, brought about through the spirited efforts of Dr. Andy Phillips of the Zoological Society of San Diego and the ministry, was well underway. And Archie was not in jail; he had been seen in Sesfontein two weeks earlier. Rumors were always difficult to verify, and there was no sense getting agitated about Archie until we knew something with certainty.

Besides us, two other Americans were delighted with the news about Archie. Dr. Franz Camenzind and Chip Houseman, filmmakers under contract with ABC Television, had arrived from Wyoming. Franz had contacted me more than a year earlier and now he and Chip planned to be with us for several months, hoping to bring the plight of rhinos firsthand to American audiences. Carol and I worried what impact these two would have on our lives and data. They seemed receptive to our schedule and agreed to be totally self-sufficient. Sonja loved the idea: Now there would be more than just Mom, Dad, and Archie for entertainment.

Before leaving Windhoek I received a shocking message from my departmental chairman back at the University of Nevada. He had been contacted by the U.S. State Department and U.S. Customs. The operator at the Safari Hotel handed me scribbled numbers and said I must call them. *It has to be about the night vision equipment.* I wanted to pretend that I hadn't gotten the message, that we'd already left for the field. Instead, I phoned.

"Hi, this is Joel Berger calling from Africa. I received a message to call this number."

"Just a second," said a woman; her voice was quickly followed by that of a man.

". . . do not mention the equipment by name. Your life may be in danger . . . people will waste you to get it, especially to the south . . . it must be returned immediately."

To the south, how could that be? Do they mean to the north, Angola? Yes, Angola would make more sense. No, U.S. intelligence is ordinarily very good. At least I hope it's good. But how could they be so upset about the night vision equipment? I phoned, faxed, and applied for permits. I did everything humanly possible to be up-front. The week after we arrived in Namibia we even visited the U.S. embassy and explained what we had done. Embassy personnel didn't think it was a problem.

Now, I was told, the university had been fined for illegal shipment of classified goods, a violation of federal law. *The university will be overjoyed with that.* I asked how something that was purchased by mail from Redmond, Washington, could be illegal, and I said that I'd appeal. Before I'd finished my thought, the customs agent interrupted, "You cannot appeal. The penalty is final."

"But I have rights. I cannot be denied due process," I stormed.

Later I agreed to return the equipment for a less sophisticated version. The university was never fined.

Errands in Windhoek were almost completed. Sonja had avoided the daily boredom of parental chores by playing with her friend Catherine at the Barnard/Simmons home. Soon we'd be in the field, finding rhinos, determining who was born and who had died, and continuing to gather data on ecology and behavior. A spate of blown fuses, negotiations over the cost of new 12-ply tires, and an overheating Land Rover were only temporary deterrents.

Finally, three and a half weeks after fleeing the cold and snow of Nevada, we were out in the real Africa. Large birds, a jumble of bouncing feathers, jockeyed for access to a giraffe carcass. With long probing bills and thick white breasts, Marabou storks feasted on the carrion. Probing the ground nearby was a kori bustard, Africa's heaviest flying bird. The bustard's remarkable capacity to consume the gummy exudate of acacias had led to its Afrikaans name, Gompu.

Driving was one thing, getting somewhere was quite another. During our dizzying stay in the States, we had forgotten that the notion of hurrying was foreign in this part of the world.

Time in Africa is nonurban, generous. In each area we traveled, we stopped, said hello, and learned of goings-on while we'd been away. Tea enjoyed under tall shady trees and talk of drought, politics, and conservation were welcome diversions. With communications limited by vast distances, the only real avenue for information was visiting. Along the way we found the "Impala People" again, Aron and Wendy, and caught up on their project. Plans had finally been approved to capture a small number of Etosha's black-faced impalas and return them, wearing radio-collars, to historic sites along the Kunene River. The bad news was that the dates for the translocation exercise still hadn't been scheduled.

When we arrived in Sesfontein, Archie was waiting. We had exchanged letters from the States, and he had agreed to work for us with the film crew along. He admitted to anxiety, but not about Franz and Chip. The thought that we might abandon him had nagged. Previously he had heard idle promises from foreigners, assurances that were never backed up by action.

During the prior year, Archie had said, "Joel, if you hear that I am not around or have found another job, do not believe it until you hear it from me. Too many people tell lies."

In return I offered the same pact: "If people tell you that we've hired someone else, or that we don't want to work with you, or that we aren't coming back, don't believe it unless I have told you. You can work for us as long as we have money and a project in Damaraland."

Archie had not forgotten. Now, he and Herero smiled at our arrival. Sonja played with their daughters, Maiyola and Fabiola.

Dense green Salvadora bushes colored the otherwise gray wash near the Uniab River. We stopped to rest briefly from the long bouncing drive. This was our first day in the field with the film crew. But no sooner had we squeezed onto the sand shaded by the Land Rovers, a rhino rocketed out of the wash, hardly 100 meters from the vehicles. It was *Kokerboom*, the young female that had treed Archie and Carol back in 1991. Within an hour we'd photographed her, and Franz and Chip had filmed a sequence that went flawlessly. We even retrieved fresh dung, packing it into a thimble-sized vial, sealing it, and tucking it into the refrigerator. Later analyses of fecal progesterone would be useful for determining pregnancy status. We marveled at the luck of stopping so close to a resting rhino.

"Three rhinos at one stop, not bad!" said Chip.

Cunningham and I exchanged puzzled glances and then looked at Archie. What did he mean, *three* rhinos?

"Two rhinos ran up the hill behind you," Chip added, pointing to a small ridge.

Archie and I trotted over. Sure enough, fresh tracks dotted the wash, up and over the hill, and toward distant euphorbia-cloaked mounds and plains. Two hours later we returned with more close-up photos.

During the past few weeks, while organizing supplies in Windhoek, the film-makers had wearied of our accounts of endurance and endless days with no animals. Now, on their very first day, three rhinos. We could read Franz and Chip's minds: This was going to be easy!

The ecstasy was short-lived. The very next day we covered 92 excruciating kilometers in 100-degree heat and saw nothing, not a single animal. But the third day we got two more rhinos. Already our total was five. Excellent.

The rain last season had been poor, and the veld now looked very dry. For days, clouds built and darkened the skies to the east. The smell of rain tantalized thirsty animals and us. Evenings, we told stories, or just sat, listening to leaves rustling in a gentle wind or a crackling fire.

One night when our mouths were parched from a day with too little liquid, I grabbed the flashlight to find my way to the Land Rover for drinks. Something moved. I lit the ground in front of Carol's outstretched legs just in time to see a scorpion. Carol whisked her legs away, while the rest of us crowded in for a better look. The arthropod scurried toward darker ground in a valiant attempt to escape, first hissing, then clicking and clacking. Carol pointed out to Sonja its essential features—the needlelike, stinging barb, the pincerlike claws used for grasping prey, and the elongate tail, really an abdomen.

Scorpions have existed for some 400 million years, but, despite their frightening appearance, most are not dangerous. Of 1,500 species worldwide, less than 30 have neurotoxins harmful to humans. Although the venom has been compared to that of cobras and can cause paralysis, convulsions, and cardiac failure, it rarely kills people. Baboons and meercats break off the stingers before eating them. Mongooses are immune.

Despite our interest, none of us wanted to risk getting hurt. Stories abound of desert travelers being stung by scorpions as they pulled on their boots in the early morning. We were cautious. Sonja knew not to lift rocks or sticks and recognized the differences among spiders, crickets, ants, beetles, ticks, and snakes. She was rarely unsupervised. But the possibility always existed of uncovering one of these fascinating predators, when either gathering wood for cooking or sitting around at night.

In addition to scorpion stories, desert dwellers delight in recounting tales of other species. One piece of folklore tells of the propensity of rhinos to stamp out campfires. We never knew whether such stories were true, but three times rhinos approached our Damaraland camps at night. All were at the same site but in different years. In fact, because of darkness and because we had been sleeping, we wondered whether the trills and snorts had been in our dreams. Only the next morning, when we saw fresh tracks, had we known that we weren't mistaken. After following the spoor, we had discovered that all the individuals were different.

The most excitement came one windless evening as Carol was reading aloud to Sonja inside the Land Rover, trying to settle her for the night. I read by the fire. Archie was at his tent. When an explosive breath alerted that a rhino was close by, Archie shot like an arrow toward me.

"Get up [on the car], Joel! This rhino can be very dangerous."

I heard fear in Archie's voice. He launched himself onto the ladder on the back of the vehicle. Crouching, I inched toward the rhino, hoping for a positive identification.

Behind me Cunningham yelled, "What's going on? What are you guys doing? Stop rocking the car." Annoyed by the commotion, she didn't want Sonja distracted from sleep.

The rhino fled. Carol had been completely unaware that a rhino had just walked into camp. She'd thought we'd been rocking the car for a diversion, silly boys with a new game.

We tracked several more rhinos. Franz and Chip didn't understand the hoopla when Carol, Archie, and I were ecstatic about our success, remembering our many rhinoless days of prior years. Unfortunately, our current problems were equipment-related. The short-wave radio stopped working, as did our fluorescent light, a 12-volt battery pack, and a handyman jack. The refrigerator's electrical system worked only sporadically, and plastic containers made brittle by the relentless sun cracked and spilled gallons of water down the side of the Land Rover.

Most of all, though, we worried about our engine, which had begun to overheat regularly, and wires that had smoked under the Land Rover's bonnet. On our eighteenth day in Damaraland, we aborted fieldwork, needing a Land Rover specialist, and drove to Swakopmund on the Atlantic Ocean. On the way, we stopped to visit the officials at Skeleton Coast Park. Rudi Loutit, who we hadn't yet seen this year, happened to be there. I smiled, then asked if he had received the notes I had sent from Nevada. He didn't respond. I extended my hand.

"Rudi, hi. How are you?"

He glared, then looked the other way, refusing to talk to me, although he exchanged pleasantries with Carol and Archie. All we could figure out was that the piece published in *Nature* had upset him since it hadn't given unequivocal support to dehorning. Perhaps it was simply a bad day to say hello.

At Knobloch's Garage in Swakopmund, the owner, Mr. Knobloch, tried to diagnose our Land Rover problem.

"That's an old illness. It's complicated. I'll see what I can do."

We liked the Namibian sense of humor—expressive with a minimum of verbiage. For three days we waited in the coastal town.

Archie waited with us. The traditional way of dealing with black workers or laborers was to drop them somewhere. If they had to travel with you to a place like Swakopmund, they were left at the rental cottages for fishermen. Reeking of fish and bait, these hovels were available for only $10 a night.

We chose a European-style hotel a block from the ocean, the Europa Hof. It had a restaurant and 24-hour parking security. Being in the bush had its uncertainties, but these paled compared to life in the cities. Even in Windhoek and Swakopmund, car thefts and break-ins were legion. Franz and Chip agreed that the protection at the Europa Hof was well worth the cost.

We entered a room furnished with bed sheets, a duvet, and a radio. There was a blue-tiled bathroom with a flush toilet, running water, and a huge spotless mirror. Archie turned to us.

"Where is my room?"

"This is it." Archie's eyes grew large. A smile came over his face.

"Mine? I've never been to a place like this."

During the three days that Knobloch worked on the Land Rover, we walked on the beach, ate in restaurants, and visited pubs. At a local bookstore Archie was questioned by a suspicious clerk.

"What do you need?"

Archie said nothing. He looked down the adjacent isle toward me and then stared at the floor. I didn't know what to think. *Is the response because he's the only black in the store? Is his unattended browsing making the white clerk uncomfortable?* We left.

Then, Archie, Carol, Sonja, and I went to see *Far and Away,* a silly American film about a poor Irish immigrant's love for an upper-class woman and their adventures settling in the Oklahoma Territory during the 1890s. Seeing a movie highlighted the dilemma we were experiencing. Archie was caught in the middle— that gray zone somewhere between the glitz of capitalism and western materialism on the one hand and life in his home town, Sesfontein, on the other. He was our friend and coworker. By including him we couldn't help but wonder if we weren't reenforcing things he could never have.

At the garage, Knobloch wasn't convinced that our Land Rover had an over-heating problem, suggesting instead that the deserts must have been unusually hot. We couldn't disagree.

Two Land Rovers, ours and the filmmakers', left the cool coastline behind and were absorbed back into the desert's shimmer.

21

Trails of Dust

[Carol]

We were having trouble finding rhinos again. During the past three days the Land Rovers had crawled more than 300 kilometers, and we had seen only one springbok and four gemsbok. The dramatic absence of game suggested that most of the large mammals had moved to other areas to find food. We broke camp and moved another 50 kilometers. Perhaps a different river might have more life, including rhinos.

Sonja sat in her usual spot, on the storage box between the two front seats, jabbering, and instigating fights between her stuffed zebra and another favorite doll, Bert. She seemed unaffected by the glaring sun and monotonous drive. Clambering to the top of a hill we saw what looked like a mirage, water shining in a rocky saddle between two low hills. It appeared before us, wide and deep, several hundred meters from the road. Trails led there, paths worn by hooves. It was real. Rains so far west were always unpredictable, but this depression had captured a good shower.

Joel grabbed Sonja. In they went, splashing and laughing. I joined them, and we floated in water half a meter deep. Springbok feces bobbed on the surface, like large chocolate chips. Franz and Chip joined us. Archie sat on the shore soaking his feet. Despite the 104°F heat, our brief interlude enhanced the rest of the day's journey.

Our new camp offered rugged canyons and desolation for 50 kilometers in every direction. A few quiver trees and the deadliest of the euphorbias, *Euphorbia virosa,* were the only conspicuous plants. Latex from this cactus lookalike (gifboom in Afrikaans, or poison tree) had been a toxic part of the arsenal of early hunters. We avoided touching any of the euphorbs after learning how one man had been hospitalized with burning genitals after touching the plant accidentally and then urinating. Rhinos were immune and, amazingly, ate the different euphorbias.

We named this place "Windy Pass" after a memorable first night. Powerful gusts bowed Sonja's tent walls, forcing them down toward her face. We traded tents. The walls flapped and swayed. Franz and Chip were also having trouble. Archie fared best. He had wisely placed his tent on the hill's lee side. No one could sleep, and we finally moved the cars to block the gale.

A few more days and we found some fresh rhino spoor at last. We had also been watching the Land Rover's temperature gauge creep into the red zone. Back in Swakopmund, Knoblach had diagnosed our car problem as a faulty gauge that read hotter than the engine actually was. We weren't convinced and watched the gauge warily. The last thing we needed was to fry the engine out here. At least Chip and Franz's Land Rover was working fine.

One morning, when everyone else was out following rhino spoor, invading ticks forced Sonja and me to abandon the shady ground under the trees and climb into the Land Rover. Massing like sharks for a feeding frenzy, they streamed from the earth, unsuccessfully trying to scale the tires to reach our warm-blooded bodies. Sonja splashed in her new red tub on the back seat as the heat surged to 110°F. Droplets trickled from a wet cloth down my chest and back.

It was day 24 since leaving Okaukuejo. Empty clouds tortured us. We hoped for rhinos and rain. I was lethargic and tried to focus on our good luck, 14 rhinos so far. But the weeks were dragging on. I wasn't sure how to counter my bad attitude. Our 12-hour, 103-kilometer drive to Doros Crater had not helped. There weren't even any songbirds.

Lightning illuminated the evening's eastern sky, perhaps 50 kilometers away. Thunder was sporadic. Calm and clear above us, stars shone. There was no wind. We all wondered what it was like to the east, up in the mountainous escarpment. That's where the rain had to be. When we had driven over the pass recently, the surrounding hills were green, thick with elephant sign. There were mountain zebras, and kudus, and springbok. Why rhinos didn't migrate but remained in the desert, we didn't know.

The next day we drove in the dry Doros River. Where we had customarily seen giraffes, gemsbok, and ostriches, there were none. In the Guantajab River it was the same story—nothing. Finally the sun peaked over the canyon's wall, heating the morning's relatively cool air—86°F. A glaring light caught my attention—reflected from the surface of a distant rock, I thought. Almost at the same time we realized it was moving, coming straight toward us. It was the leading edge of a muddy torrent, undoubtedly from last night's ferocious storm. We needed to move to higher ground and in a hurry.

Flash floods are renowned in these canyons. In his 1855 journal, *Travels in the Interior of Africa,* James Chapman wrote of the Namib Desert: "all the periodical rivers hold water from 6 to 12 inches below their beds of sand and gravel. When rains fall 200 miles east in summer, overwhelming floods roar down suddenly and often carry off some unfortunate family sleeping on the bank."

Among the ill-fated was Garth Owen-Smith, who once lost a Land Rover this way. We quickly moved and parked up and away from the closest bank. This tributary to the Ugab could flow for days. Almost always dry, these rivers ran once or twice every 10 years, according to Archie. He'd never seen a leading edge before.

We jumped out of the vehicle, Sonja and JB sliding on the slick shoreline. Gravel bars appeared. The murmur of rushing water filled our ears. Then someone shouted, "Water fight."

In a moment, thick mud clung to my T-shirt. Dirty water sprayed my face and chest and coated my legs. Franz and Chip grabbed cameras and sprinted to film the leading edge as it moved down the riverbed. After an hour it was time to continue working.

Up ahead, the river channel narrowed to a deep gorge. With water, it would be impassable. We looked for a route away from the rushing current.

Twenty minutes later, when we climbed onto dry land, we saw a dust plume rising from the adjoining hills. Four hundred meters away, a Toyota Hilux approached at high speed. In the bed of the yellow bakkie stood a man. When it was 200 meters from us, the driver braked abruptly and then cut away full tilt across virgin desert, leaving behind a trail of dust.

We tried to assess our options, thinking about a high-speed chase, guns, police, jurisdiction. We didn't know what to do. We kept thinking, "We're not enforcers, we're here to find rhinos." What about Soni? They must have guns. Why else would they race away from us like that? Minutes passed with more discussion. They were getting farther and farther away. The time for action had expired. If we had wanted to follow, we should have done so immediately. Our respect for rangers and police, people who must evaluate situations quickly and act on snap judgments, increased.

We decided to backtrack the men from the truck's spoor to see what they had been doing. Their behavior was strange, unnatural. They had raced away as if guilty of something. Since we began in the desert in 1991, we had only met vehicles five times, and people almost always had stopped. Something wasn't right.

The tracks meandered aimlessly across roadless land—river bottoms, valleys, plateaus—areas where kudus, gemsbok, and zebras could be found. We followed the tracks for 12 kilometers. Their travels appeared random. Nothing made sense. All we could assume was that they were looking for meat and, after seeing us, panicked and fled.

About 90 minutes before sunset, a series of prints 20 centimeters wide appeared in front of the Land Rover. A rhino had been walking from Doros Crater due north, mostly on the road. Black- and yellow-winged butterflies probed its soft dung. This rhino had been here recently. We followed the spoor for four kilometers until it veered across a nameless valley and toward distant hills. If we wanted photos, we'd have to hurry. Archie and JB went out to track the rhino, followed by Franz and Chip.

Nearly two hours passed. Finally Soni and I saw the four men shuffling back across the stony plain. Sonja ran toward them in the fading light. Joel told us what had happened.

"At 100 meters the male didn't see or smell us. When I was 50 meters from him, I began taking pictures. Franz and Chip were filming. Everything was fine, but within seconds there was a full charge, completely my fault: I still needed a facial shot, and I snapped my fingers to get him to look at me. I didn't even get the photo, but I saw his ears. It was *Sandy*. Then I was up in the tree. There was only one thorn tree for safety. Franz was close behind me, yelling 'I want to climb up your butt.' I didn't know where Chip or Archie were."

The attacking rhino had almost horned Franz, but somehow he'd escaped. Archie's shirt was shredded from thorns, his back and shoulders full of lacerations. JB had 23 punctures, and blood oozed down his leg.

Twenty-four hours had made a striking difference in our dispositions. We happily downed unheated soup at 9:30 P.M. The temperature cooled to 92°F. *Sandy* had been the first rhino dehorned in Doros Crater, and no one had seen him in more than two years. We now had horn regrowth data for another animal. We climbed to our bed under the stars and watched lightning illuminate the mesas far to the east.

Everything we owned now had a place. Despite the apparent chaos inside our Land Rover, Chip and Franz were impressed. We could find equipment when we needed it. The Land Rover may not have looked like an organized home, but it was.

My front seat held the cutting board, a Swiss Army knife, magazines, and two pairs of binoculars. On the floor were dolls, modeling clay for Soni, and peanut butter. Music tapes, three water bottles, and maps were on the shelf and the dash on my side.

On the driver's side was another Swiss army knife. A whistle and a tiny thermometer hung from the turn signal switch. Two visors swung from the mirror.

Inside the chest, which Sonja often sat on, were other supplies—ethanol (for dung samples), the invertor that changed 12-volt current to 110 current, data books, and loose screws, nuts, and bolts. Books on emergency medicine, local plants, mammals, birds, and snakes made up the lower layer.

We kept scientific and photographic equipment on the back seat behind the driver—rangefinder, monopod, lenses, more data books, film, and small packs, all snugly tucked into Sonja's red tub. Between my seat and the back seat we stored Sonja's straw book-basket, more music tapes, a box of Sonja's cassettes, and her rarely-used hairbrush. Twenty-liter water jugs, tent poles, and straps and cables filled the rest of the leg space. The fluorescent light, medical kits, rolls of toilet paper, tire irons, a lug wrench, and the crow bar were buried under the seat.

Fish netting drooped from roof hooks over the rear storage compartment. From within, among foil, paper towels, and clothing, peered Sonja's stuffed zebra and assorted dolls—Ted, Bert, and Softie. Below, there was a tub that kept fruit from being crushed and held water when we washed dishes.

The back held the refrigerator and soft clothing bags, inflatable seats (for sitting in camp), hard briefcases, a large camera case, a Samsonite case holding the solar panels and computer, wires and electrical gear, boxes of canned food and juices, Soni's yellow ball, hoses, clamps, oil and lubricants for the car, the Cadak camp stove, two 10-liter water jugs, thermoses, a red backpack for toiletries, and a cast-iron pot.

On top of the Land Rover were four to six jerry cans, four water jugs, three tires, a shovel, the high-lift jack, and three full trunks. A backpack was strapped to the top of each trunk, one holding JB's clothes, another holding tents inside a duffle bag, the third holding bedding, sleeping bags, netting, and tarps.

We disliked breaking camp, packing and unpacking, but gradually we did it more efficiently. Back home in the States, living this way wasn't comprehensible. Camping for a few days, a week or two at most, was one thing. Living out of a car so that we could track rhinos, while trying to be good parents and carrying on a research project, was outrageous. But if we wanted data, there was no alternative. There simply wasn't an easy way to study rhinos, at least not in this part of Africa. And after doing it in 1991 and 1992, it seemed natural. By 1993 we couldn't imagine any other lifestyle. We had become gypsies trailing dust.

22

Of Moths and Maggots

[Joel]

I woke to yellow and blue, early morning through colored nylon. Our cocoon of warm air deflated as I lifted the sheet and crawled away from Carol and out of the tent. Thirsty, I remembered there was Naartje juice in the Land Rover. Already tasting the tangerine flavor, I swung open the back door. A flurry of dusty black bodies and flashing yellow eye spots dashed toward me as moths exploded from the car. Hundred of wings beat wildly into my face. My eyes squeezed shut, and I flailed at them with my hands. Quiet Archie grinned as he sat by the fire.

For weeks, the moths erupted. "They come with the rains," Archie said. Some days we barely noticed a few, other days there were thousands. Moths clogged containers, landed in food, and swarmed over plates. Detached wings and decomposing bodies oozed into the water-filled dinner dishes from the night before. Moths materialized under seats and in the crease where the dashboard met the windshield. A few crept into juice containers and later slid against our unsuspecting lips.

Mopane moths, also known as emperors, enjoy a precarious existence. They hatch at six millimeters and eat their empty egg shells for a first meal. By their sixth molt, the brilliant red-yellow-green-and-black worms are seven centimeters long. Those that haven't been killed by parasites, bats, or birds face a new threat— hungry humans. From Mozambique and Zimbabwe to South Africa and Namibia, mopane worms are highly sought-after treats.

"You pull their heads off. We eat them uncooked, but they taste better roasted over the fire," said Archie, remembering his boyhood.

I imagined plump wriggling bodies in my hands. Instead, I asked, "Are they as good as roast goat?"

"Well. Many people like mopane worms, but goat is my favorite." I half-expected a wink.

One day in the hills above Sesfontein, Soni, Carol and I hunted mopane worms.

"Dad, over here." Sonja found them, guarding against runaways until Carol or I could pick them up. When I looked again, the colander for tonight's noodles was the new home for six luminous worms. Cupped hands blocked their escape. I still couldn't imagine biting off their heads.

"Sonja, do you want to help me let them go? I bet we can find a nice tree and put them back on the leaves."

"Sure, Dad. Can you help me climb?" She was already scrambling up the trunk when I reached her.

Although the idea of eating insects, cooked or uncooked, is abhorrent to many westerners, grubs and other invertebrates have been human food ever since we humans evolved. Baboons and other primates also savor insects, which are full of valuable protein and vitamins. A single dried caterpillar weighs less than six grams. One hundred grams of mopane worms offers the average person their minimum daily requirement of calcium, riboflavin, phosphorus, and iron. So if a single dried caterpillar weighs just less than six grams, a person only needs 16 worms or so to meet the daily requirement—a trivial number by most standards, since experienced laborers have picked as many as 16,000 grams (nearly 2,600 worms, or 35 pounds) in an hour. Even by hand, huge harvests are possible.

This is not just a rural phenomenon. Mopane worms are marketed in the cities. On the streets of Johannesburg bags of dried worm morsels are as popular as the canned version, mixed in tomato sauce with chili powder.

* * *

Meanwhile, the international media appreciated a good story. Dehorning headlined countless newspapers and magazines. Articles entitled "Cutting the Rhino's Losses," "Giant Pig with Number on Back," "Spiking the Guns," "Dehorning Rhinos to Give Them a Future," or simply "Save the Rhino" filled the pages of *National Geographic, World, Leadership,* and *Africa South* magazines.

Dehorning had been initiated because, despite 25 years of formal protection, conventional tactics were not working, and southern Africa refused to accept its rhino losses. Countries decided that, because poachers continued to kill rhinos, a new program was needed. The horns must be kept from the poachers. If they tracked rhinos in Namibia, they would find the animals hornless and without value.

The plan was to protect the "removed" horns in vaults along with those confiscated from poachers and those sheared from animals that had died naturally. Later, when the international ban on horn trade would be lifted, the ministry would sell the stockpiled horns. Money from these sales by the government itself could pay for future conservation efforts.

However, the ban blocks the process: Rhino horn cannot be sold legally, even by host governments. The Convention for International Trade of Endangered Species, to which there are more than 100 signatory nations, prohibits the sale or trade of rhino horn. The last time CITES had met, the policy of horn prohibition had remained in place.

Another variation on the issue made headlines, the question of whether dehorning works. The optimistic view found in media releases set the tone: "So far, none of the dehorned rhinos in Hwange [Zimbabwe] or Namibia have been poached" or "the dehorned animals have been spotted subsequently, mixing with beasts still with their horns, and proving that dehorning doesn't confer any kind of 'apartheid' status in rhino society." In the September 1991 issue of *Observer Magazine,* three Namibian calves were reported as being born to dehorned mothers, but an article in the January 1992 issue of *National Geographic* implied that Blythe Loutit had recorded only two. When we asked her about the identities, she couldn't remember who the mothers or calves were. Although such details worried us, the world's community of rhino watchers were encouraged by these reports.

Now back in Etosha, we returned to our routine. Carol worked in the darkroom, I analyzed data and answered correspondence. Sonja played with her pals, Pieter and William. We also visited biologists and other friends at the Ecological Institute. At night, we sat at the floodlit water hole hoping for rhinos.

Late most afternoons we were back at the airstrip jogging, and relishing the game-filled horizons that had been lacking in the drought-ridden Namib. One spectacular early evening 1,500 springbok, 300 zebras, and a dozen wildebeest watched as we ran along the runways of short grass and earth.

As I walked the last 100 meters in a good sweat toward the Land Rover, Sonja announced, "I have a big, *big* surprise."

"Is it alive?" I asked as I grabbed a water bottle.

"Yes"

"It's not a plant?"

"No."

"Is it nice or nasty?" I asked, thinking that they must have found a snake curled nearby.

Sonja decided she'd heard enough questions and raced to the car.

"Look between the door and the seat, Dad," she said, giggling as she pointed to the floor. "There are sooo many."

She was right. Bits of moth bodies, drying but still moist, had fallen there, creating a veritable warehouse of nutrition for hungry maggots. Compared to the few harmless lentils that had sprouted from the carpet last year, the numbers of tiny wriggling fly larvae were enormous. Fortunately, they were not immune to silicone lubricant spray and suffocated under repeated filmy coats. Cunningham and I wondered how this reflected on our living habits or conscientiousness as parents, or both. We agreed to tell no one, hoping that Sonja would soon forget and not blurt something out in public.

"No more maggots, Dad," she'd declare occasionally, reminding us of her impressionable mind.

A week later we caught up to Bill and Kathy Gasaway. Over dinner, they updated us on their project studying predators and prey. The operation had grown. Ministry staff and other project participants were estimating ungulate predation rates by radio-collaring cheetahs and jackals and collecting data on ungulate body condition and causes of death. Volunteers from the Netherlands, Malawi, the United States, and South Africa had joined the Namibian staff in an attempt to meld conservation and science, management and wildlife. Bill and Kathy were exhausted. They struggled to communicate to the States from the faxless and sometimes phoneless Institute. Coordinating logistics with the USAID mission and the ministry offices in Windhoek, and arranging for equipment and funds through the Zoological Society of San Diego back in California, as well as conducting field research (sometimes for 36 hours without sleep), was so stressful that Kathy occasionally wished the project had never been conceived. We hoped a visit from us was still welcome.

A menagerie of tall and short camping chairs surrounded the fire. We sat and relaxed, tantalized by the aroma of Bill's baking bread. Then my curiosity about the maggot eruption overwhelmed embarrassment. I slipped in a comment about maggots, edging the conversation toward decomposition and life. Amazingly, maggots had also infiltrated Bill and Kathy's camp. Theirs were in the drain below a refrigerator where dead moths had fallen. Since our moth buildup was similar, we confided our secret and thought that perhaps we weren't so unfit to be parents.

Kathy agreed. Others might think we should stiffen our laissez-faire attitudes, but after all, what were a few maggots on the floorboard? Since we weren't likely to find too many people in agreement, we'd still keep our maggot affair a secret.

* * *

We planned another trip into the northern Kaokoveld, where last year's rumors of persisting rhinos continued. We'd also check along the Kunene River, the shared boundary with Angola, and look for our Himba acquaintances. But the Land Rover's clutch problems and overheating continued. We decided that, before meeting Franz, Chip, and Janie (Franz's wife, who is a still photographer and sound recordist in Damaraland, we'd detour to Outjo and Weiman's Garage.

By the time we arrived, Weiman's was already closed for the day. Our Land Rover checkup would have to wait until the morning. We stayed at the Onduri Hotel, leaving the fully loaded Land Rover in the fenced parking area. An enormous crash of lightning and thunder woke us at 11 P.M. A downpour was drenching the mattresses, duffle bags, and bedding still on the roof rack, and we hurried outside to retrieve them. Back in the room, we found we'd been joined by about a half dozen buzzing mosquitoes.

Then we couldn't sleep. At 2 A.M. I began sneezing uncontrollably. With each expulsion, my head burned. Carol watched helplessly as sweat covered my body during the 40-minute attack. Security lights outside penetrated the sheer curtains and lit the room. We turned over and over, searching for a comfortable position. My exhaustion and frustration grew.

Suddenly, Carol grinned.

"I know what's wrong," she whispered. "We can't sleep because we're on the wrong sides of the bed!"

I stared speechless. My mind recreated our sleeping positions in tents, on the Land Rover's roof rack, and in our Nevada bedroom. She was right. We were on the wrong sides of the stupid bed. We switched.

On familiar territory, the tension melted away. We slept. Sonja slept through everything, the thunder, the rain, the light, and the sneezing. Our sheets and blankets hung in the bathroom, dripping from the storm.

In the morning, Weiman's mechanics fiddled with the car, but the overheating baffled them. There didn't seem to be anything to fix.

Driving north from Sesfontein toward Angola, traces of the twentieth century grew sparse. At Purros, several Herero woman crowded around a fire. Their dogs eyed us warily. Two bare-breasted Himbas shot quick glances, then moved back toward a crying baby. They were traditionally clad, with red ochre smeared on their skin, bead necklaces, freckled creamy conch shells, and many bracelets rising from their wrists, all above brief skirts of tanned skins.

In this part of Africa, Purros has grown famous as one of the few areas where Himbas still live traditionally and can be seen by tourists who don't have four-wheel drive. This day, most of the Himbas had gone into the mountains or canyons, testimony of their indifference to anthropologically bent tourists.

The two-track passed the refreshing paradise of the Hoarisib River, dry but vegetated and shady, before reaching plains still green from recent rains, on its way to stoney deserts. Near the Etendeka Mountains, ostriches shot up and ran, soon fading like spirits into shimmering haze. At the Hartmannberge uplift, a chameleon slipped into a shady burrow. An agama pressed into a rocky crevice. Then we dropped into a valley, 100 kilometers long, filled with rugged outcrops, steep walls, and hidden basins. Grass grew deep. The rains had been good here. Springbok as thick as mopane flies pronked as they ran from us, their streaks of white and dark brown bobbing up and down in frenzied motion. The tracks of spotted hyenas crisscrossed mud. Even a cheetah had come this way. To the east, the Otjihipa Mountains stood like monuments, their escarpment rising a full kilometer and a half above the silent valley, dwarfing us and everything else in it.

In 350 kilometers we had seen only three groups of people, all Himbas. We were searching for Muhinena Tjambiru, the lean muscular leader of a small group we knew from last year's trip. He had agreed to guide us if we still wanted to cross the Kunene River into Angola. For a day we drove back and forth, hoping to find signs of recent human presence among the few mud and dung homes that remained standing. No one was there, only a few discarded baskets and beads. The people had moved on.

Our search continued. As we bounced over another ridge, a river suddenly appeared below us. Unlike the dry rivers we'd seen for months, this one ran wide and fast. Slick, black rocks converged in a gorge. White water shot by. Palms lined the opposite shore. Only the burning sand separated us from a refreshing bath. Sonja interrupted our reverie.

"Mom-mee, is that the Kunene River? Are there crocodiles?" She obviously remembered our previous trip and the crocs of the Caprivi and Zimbabwe.

Inching our way on foot down jagged rocks, we watched for snakes where the vegetation has grown dense. At a shallow pool beyond the rushing river's edge, we stepped into knee-deep heaven, and slid lower, up to our necks in the invigorating water.

Archie had never seen a real, permanent river.

"No one in Sesfontein will believe me. Can you take a picture?"

Suddenly Archie charged out of the pool.

"Something bit my foot!" he yelped.

Before he finished his sentence, I lurched out, closely followed by Carol and Sonja.

We looked at the tiny eddy, water not more than 30 centimeters deep; the torrent was blocked by boulders. There couldn't be a crocodile. Slowly, our courage returned. I slid in, stretched my legs, and lay down again. Soon there was nibbling on my toes. This time we realized that fish were mouthing us, tiny fish. We all laughed as I rose to get the camera.

After our soak, we explored the hill country, the Land Rover straining forward in low gear. Franz's Rover followed faithfully. In front of us, two donkeys, a dog, and a Himba woman and young teenaged girl appeared on the two-track spoor. They must have heard our cars struggling toward them, but they displayed no emotion and kept walking. Their donkeys carried wood. Their camp must be close, I thought.

Archie talked with them in Herero. From the car, Carol and I wondered if we'd met them last year. We weren't sure. But when Carol and Sonja stepped out, the woman remembered the "little girl with light hair." She became friendly and willing to visit, even pulling out a pipe stuffed with a wad of home-grown tobacco. Carol and Sonja were invited to join her on the shadeless gravel. Archie translated. I was the superfluous male.

The woman wanted to know how Sonja and Carol were, and where they had been. After a while the woman said she was hungry and had far to walk before the sun set. She and the shy girl disappeared over the next ridge, weighted by a few tins of fruit and fish we'd given them.

We renewed our search, trying one encampment after another. Having no success, we finally abandoned the Kunene River and tried the battered dwellings that sat on ancient terraced slopes. Along the flood plains there were signs of old crops. But despite the signs of recent habitation, all the sites were deserted. We had been foolish to expect otherwise. Some Himbas were seminomadic, especially in this remote part of Africa where the twentieth century was just beginning.

Everyone grew impatient. Franz and Chip desperately wanted a human dimension for their rhino film, and to depict local attitudes toward wildlife and conservation. Precious days had been wasted just getting to this remote place. Churning over a rocky spoor was exactly what we'd all been doing for months. The man we sought seemed more and more an apparition. And if we found Muhinena, he might not want to be photographed. Franz was sensitive to the question of exploitation and whether filming was appropriate. (He was also a professional— an artist in his element—seasoned by experiences in China, South America, and Alaska.) But at this point all Franz and Chip could film anyway was scenery. People were rare.

Near a stand of isolated palms, smoke drifted into the deep blue sky. A man wearing long baggy pants and a button-down shirt stepped forward. Archie spoke to him in yet another language, Owambo. The man told us that the people from last year had moved on. He didn't know where—maybe "Ongola," maybe the mountains—and he pointed to the rugged spine of the roadless Otjihipa Mountains. If Muhinena was up there, Archie and I would be forced to strap on packs to search for him, leaving the others behind.

The previous year we had asked the ministry about wildlife surveys in Angola. Because of the war and the potential danger, all attempts to mount an expedition had been canceled. The concept of international peace parks had increasing appeal, and the idea of connecting large swaths of land in adjacent countries was now common in talk among conservationists. If we could substantiate the existence of rhinos in Angola, perhaps the international community would be willing to support conservation there. Although we weren't prepared to mount an expedition into Angola just yet, we needed to see Muhinena. He had planned to do reconnaissance there during the time we spent in the States, and it was our responsibility to find him and follow up.

Angola's remote southwestern deserts had been surveyed 25 years earlier by Dr. Brian Huntley. Parque Nacional do Iona was Angola's largest park, some 16,000 square kilometers. It included sand dunes, steep canyons, grassy plains, and mountains more than 2,000 meters high, and had been home to many species, including mountain zebras, gemsbok, elephants, rhinos, lions, leopards, and black-faced impalas. Hippos had occurred upstream in the Kunene. Aardwolves and brown and spotted hyenas had probably lived there. Perhaps some wildebeest and plains zebras still existed. Even wild dogs might have survived the war. No one knew.

We accepted the Owambo's advice and drove toward a little village where a Herero who knew Muhinena lived. We bartered for his services through Archie, who again switched languages, Owambo to Herero. This man wanted sugar, tobacco, and meat. Archie talked with him until a bargain was reached.

At once the man climbed up to the top of the Land Rover, grasping duffle bags and trunks until he was perched among the jerry cans and high-lift jack, the tires and tents. He pointed away from the single two-track toward the roadless land. *There's no track that way.* I couldn't decide which way to go. The last thing we wanted was to cut new spoor across virgin grassland. Instead, we dropped into the dry river that headed in the same general direction.

The man began shouting.

"Joel, we must not go this way. He said that the river is difficult and the level plains are easier," Archie explained.

"Archie, please tell him we don't want to leave a trail for others to follow if we find the village with people." Archie looked at me, confused.

I continued, "Archie, you remember how we never like to make new trails in Damaraland. This would be worse. If we go this way and find Muhinena, other people with vehicles will see the tire spoor and follow it. There would be a road right to his village. We don't want to leave tire spoor, except where we have no choice."

"You are right. This is a good idea, I will explain."

For two more hours we plowed through heat and sand in rock-filled riverbeds and across plains. We all grew frustrated. Franz had now been in Namibia for three months. Although he'd filmed desert rhinos, most were running away from us. Definitely this footage was not the stuff of ABC Television prime-time specials in the States. He had yet to film rhinos at night or to see a lion. Now we traveled with a guide who wasn't even sure if people lived on some unnamed terrace. All we knew was that the camp was beyond the mouth of a canyon along the Otjihipa Mountains. How likely were we to find these people? Franz had persevered on wolf and panda projects, turning hardship to success. We all wanted to be optimistic, and we had no choice but to trust this guy and follow him blindly. We plowed on.

Just when we were beginning to believe we would never find the village, we saw a donkey, then cattle. Angling toward the slopes, we noticed they had steepened and were almost vertical. An ancient river had sliced through earth and rock to form a massive canyon. From its mouth, a flood plain spread. Huts materialized. People sat clustered under a tree. I switched off the engine, some 400 meters away.

We felt like invaders, total aliens. Assholes. Hypocrites. *How can we approach these people tucked away from machines and twentieth-century materialism? This is their home, their land. Why are we here? Forget rhinos. These are people! They shouldn't have to deal with the world's loss of biodiversity. They are people just trying to survive.* But then my wild thoughts slowed. *Why aren't there rhinos in the northern Kaokoveld? Maybe they'll help us understand.*

We made a plan. Our guide and Archie would approach the village. Carol, Sonja, and I would wait. The other Land Rover sat another 100 meters behind us. Uncomfortable, we hoped it was polite to walk in. Little girls and boys ran naked. Bigger children tended cattle. A small lamb was tethered to a tree, bleating. Beneath the tree, women worked and talked in the relative coolness of the shade. Archie and the guide disappeared. Cunningham and I stood at the front of the car. The sun burned overhead as we waited anxiously.

Archie reappeared. He and another man walked toward us, chatting. This man was definitely Himba.

At two meters, he stared, looking me up and down. I wasn't sure I recognized him. Archie muttered something in Herero, then turned to me.

"Joel, this is the man from last year. He is the one who said he would guide for us."

I felt stupid. He obviously knew who I was.

"Archie, tell him hello. Then ask if he has been to Angola and seen Ongava," I offered, remembering the Herero word for rhino.

Without smiling, Muhinena showed me the scar on his thigh, the result of crossing the Kunene on foot years earlier. A crocodile's handiwork was evident. Purplish marks from teeth spread across and under his thigh. He grinned, obviously enjoying my horrified look.

Archie went on, "Joel, he crossed the river on a donkey this year. He saw several elephants, one lion but no giraffes, no plains zebras. There were some mountain zebras, gemsbok, and kudus. No rhinos."

More men approached. Before they arrived, I gave Muhinena a knife, something from the States that I'd promised him the previous year. After I showed him how to unlock the blade he was impressed.

Carol and Sonja, Muhinena, Archie, and I all returned to the village, leaving the filmmakers watching silently from their car. Back in the shade, one of the women was grinding corn on a stone. She stopped pounding and spoke to Carol, who smiled, not understanding the words. We recognized her though. It was the woman with the donkeys from yesterday, the one with a pipe, the one who was hungry. She and her daughter had walked more than 20 kilometers before dark. Her pounding continued until only fine grain covered the stone.

Archie asked Muhinena about Franz, Chip, and Janie. He explained that they wanted to take pictures with special cameras and made it clear that they would stay at their car if Muhinena or the others did not want them in the village. They should say no to the request if they wanted. Much discussion followed, none of it tangible to us. Archie just listened, as did we. Twenty minutes passed.

Finally, Muhinena returned to us. Everyone in the village was comfortable with Carol, Sonja, and me remaining. We were welcome; people knew us from before— the people with the yellow-haired girl. And no one objected to a visit by the filmmakers if they would leave food for the village. Franz offered tins, sugar, and other goods. Obviously, they had uses for products from the "outside."

Other villagers now talked, visited, and laughed. I pulled a mammal field guide from my pack and showed photos of wildlife, asking which animals had been seen. The men were more interested in me, so Carol talked mostly with the women. Kids touched Sonja's hair. Archie disappeared. We were on our own.

Several women ground corn. Sonja moved toward the older one, the woman with the pipe.

"Can I help you?" Sonja asked. More people gathered. Lifting the pestle, Sonja struggled and grunted. Three times she thrust it down, each time missing the corn. She looked disappointed and ready to cry. "It's too heavy."

The women roared with laughter, cackling, bellies and breasts bouncing. Someone brought the baby lamb for Sonja to pet. Others stared or came to touch Sonja's fine blond hair. The tangled wisps blew around wildly.

Archie reappeared with another man. It was someone he knew from his fence-line days back at Hobatere. The man had grown weary of merchants and tourists and was happier in the western deserts. He explained that he had just returned from Angola, where he'd seen gemsbok, springbok, kudus, and mountain zebras. Muhinena came over and gestured to his face, rubbing his hands up and down in a game of mime. Carol translated this time.

"I think he wants a razor. Look how he's touching his face."

Instinctively, I rubbed my typical 20 to 30–day growth. "Archie, tell him if I had a razor, I'd be using it." Muhinena understood and laughed with us. There were smiles all around.

Through Archie, he explained that he'd be our guide whenever we wanted.

"What can we bring you next year? What do you need?"

There was a long pause and Archie repeated the question.

"For me?"

"For you or anyone here."

He seriously considered the question and glanced around his camp while he thought. As headman, he could ask on behalf of the whole village.

"Blankets."

"What else?"

"Nothing. I am happy. We are all happy."

We asked if they needed other kinds of supplies.

"No."

"None?" We couldn't believe that. These people were at the fringe of the world, making a go of it in the oldest desert on the planet, the Namib, and they needed nothing! I asked Archie to ask again. There must be things they used but couldn't regularly get from the local area.

"Okay, some mealie." Muhinena thought longer and added, "If we all go to Ongola, maybe bridles for the donkeys. Pills for my wives when they feel sick. And medicine for sore eyes."

"Archie, how many wives does he have?"

"These people are not like us, Joel. They have many women. I think he has two. See that young woman over there, she is his other wife, the one near the woman with a pipe," said Archie, gesturing.

Before we left, I told Archie that we weren't sure if we'd get permission to go into Angola. Being an American, I was hesitant to cross the river without appropriate papers. Muhinena would never understand, so I asked Archie to explain that we hoped to be back but we weren't sure if it would be possible.

As we drove away none of us spoke. Although it had been a good visit, we were depressed, upset at the possibility that modern living would infiltrate Himba society and hurt them in some way. We knew that modern medicines could help them, and that some aspects of the twentieth century could make their lives safer and more comfortable, but many parts of our culture were negative. It was a conflict between ancient and modern, between a simple life and a world out of whack.

Traveling to the northern Kaokoveld, we saw one fence, one in 950 kilometers. But change is imminent. A major dam has been planned that could slow the Kunene to a trickle, altering not only the river but the land beyond for the inhabitants—the Himbas, other mammals, and the ecosystem itself—forever.

23

The Zimbabwe Massacre

[Carol]

Just after dark a gray shape with a long regal horn strolled across the vlei. I had been reading to Sonja from the front seat of the Land Rover and clicked off the flashlight. *Why the Ostrich Can't Fly* would have to wait for a few minutes.

"Was that the daddy or mommy ostrich?"

"Whisper, Sonja, we don't want to scare this rhino. Maybe it's one we know." Slowly, I lifted my binoculars.

"Who is it, Mommy? Is it *Tina?*"

Before I could answer, a hairy face appeared at the window. Hanging upside down from the rack above, Joel hissed a quiet order.

"Shhh."

Nearly eight months had passed since we had last seen *Somalia.* It was the price of being in America and then visiting our other sites before returning here to Calcrete Basin. As usual, the rhino squirted urine at an old dung pile and sniffed the soil. Moments passed. Soft fleshy balls plopped onto the ground, and rear legs as adept as golf clubs propelled juicy dung backwards, leaving a message for the next passing rhino. But this rhino drank from the east, not from *Somalia's* customary spot on the west side of the pool. His ears rotated furiously. He was nervous. Maybe elephants or other rhinos were at the forest's edge.

On top of the Land Rover, I sorted through the photo files. The animal had a low stump for a rear horn, and the left ear had a slight tear. This couldn't be *Somalia*. No, maybe it was. Confused, I wondered if he could have broken his horn and torn an ear since our last visit. The front horn was a fair match. We thought about the other bits of evidence. They tantalized us.

Each rhino almost always came from the same direction—from its home range. When we considered the direction he had come from, we realized it was 35 kilometers in that direction to the nearest water. Perhaps it was an animal we didn't know. The lack of familiarity with the area, us, or even other rhinos might explain the erratic and nervous behavior. We had to shoot identification photos and confirm the animal's gender.

Given the distance that the rhino had now moved, Joel would have far to walk. He'd also have to get low enough for a good view of the genitalia. That meant kneeling or lying on the ground to peer upward into the night sky from below the animal. Otherwise, we'd never determine the sex.

Walking seemed safe. Just before dusk, a small breeding herd of 20 elephants had drunk for 20 minutes. They probably wouldn't come back. Lion roars were still several kilometers off. We couldn't know, but we hoped no lions were closer. Two other rhinos, a mother and calf, were somewhere in the nearby woodland. It was a good time to get photos.

JB approached slowly and deliberately. When he reached the second rock marker, only 35 meters away, the rhino knew something was happening. I watched helplessly from my usual position on top of the vehicle, 100 meters from Joel, one hand glued to the monopod, the other gripping the night scope. *Male or female? Males usually don't charge. Male or female? Teats or testicles?* Both of us thought of nothing else. Joel crouched. He and the rhino stared toward one another. Less than 30 meters separated them.

Methodically I scanned with the heavy night vision scope. No elephants. No lions. One far-off hyena. The mother and calf rhinos were still 250 meters from the water. I repositioned the scope toward center. Joel was kneeling, still trying to get a good look. The rhino lowered its head. If it didn't offer a rear view, we'd probably never know its sex, especially if it was a transient on a one-time visit.

Joel unholstered the .44 magnum. *What's going on?* My adrenaline rushed. I still didn't see a thing, nothing dangerous near him. He lifted the gun. Joel had never drawn a gun during the entire time we'd been working in Namibia. The rhino hadn't moved. *What the hell is he seeing?*

Instantaneously I shot 750,000 candlewatts into the rhino's eyes. They shone back. Then, it turned. A male after all. Slowly, JB returned to the car.

"What was it, Cunningham?" said Joel, looking up at me. "The rhino wasn't doing anything. Why'd you flash (the spotlight)?"

"Why'd you pull the gun? I thought the rhino was ready to charge. Better to blind it, right?"

"The rhino wasn't the problem."

"So, why was the gun out?"

"Didn't you see the jackal? It came straight at me," he said. "I thought it might be rabid. If it kept coming, I couldn't wait forever. I saw it at 30 meters. Remember, a rabid jackal at Okaukuejo bit Kallie's wife. I didn't know why the jackal was so bold."

Joel climbed onto the Land Rover and poured a cup of coffee. High-pitched mewing sounds bounced back and forth across the tranquil vlei in the darkness, the sounds of baby rhinos. At 2:30 A.M. we turned giddy.

During the next two nights, 17 rhinos came and left. *Somalia,* without torn ears but with majestic horns, approached calmly from the west—our photos confirmed it. We also saw a one-horned rhino. With its back horn snapped off at the base, it was the closest thing to a unicorn we'd ever seen.

The moon would rise 90 minutes after sunset. JB was sitting on the Land Rover, despite the total darkness. Soni and I were inside, with the windows half-open to let in the night sounds. The quick pulsing slurps of a leopard were followed by "whooop, whooop, whooop." We saw four spotted hyenas on the move.

"Poop, pooop!"

I opened the door.

Zebras stirred restlessly. I looked up at JB.

"What can I do? She has to go," I said.

"The grass is in the way, Mommy."

"Sonja, shhh, you'll disturb the animals."

"But, Mom, the grass."

I carefully held the raspy tufts away from her, and, once she was back in the car, I buried the evidence so Joel wouldn't step in it. Sonja should be able to sleep now. "Mom" was happy.

*   *   *

It was serene at Okaukuejo. Springbok and zebras grazed the open savannas. The roars of lions and the trumpeting of elephants entertained visitors. Tourism was alive and well in Namibia. The aura was misleading, though. Not all was right.

The shelves at Okaukuejo's store were more barren than ever. Petty thefts were on the increase. The postman had been charged with larceny. The head of the ministry's elite Game Capture Unit, Louis Geldenhuys, had been accused of conspiracy to sell sable antelopes. Park privileges were being abused by Etosha employees and their friends. Parties in the veld, gates left open, and vehicles on dusty roads at night fueled escalating fears of poaching. Three black rhinos had been shot in the last six months. At Waterberg Plateau to the south, seven white and at least one black rhino had been killed in a wave extending to private ranches.

Then there was our friend Bill Gasaway, whose million-dollar grant project was reaching horrendous proportions.

Bill had traded fieldwork for mountains of paper, trying to develop a scientific basis for management decisions. The management staff and Bill were effective in coordinating the ecological work, but, unlike past times when he and Kathy had combined data collection with analyses and correspondence, their time was now spent ordering equipment, meeting with Ministry staff, and visiting section chiefs. New facilities were being built to house personnel for their project. Planning and arranging for equipment and laborers took time. Always there were delays, the latest a long wait for the approval of different components of the project from offices in Harare, Windhoek, San Diego, and Washington, D.C. Volunteers required living quarters, electricity was needed in the Institute's library, and computers weren't running. On some days, vehicles needed repairs. On others, a receiver malfunctioned, cheetahs couldn't be followed, or lions had been shot outside the park. The project took on a new dimension, one which excluded the word *fun*.

Joel met Soni and me at the Okaukuejo post office. The sun was already above the concrete entry gate, glaring and hot. Receiving our mail was always a highlight, especially after we'd been out in the desert.

"So, Cunningham, I have good news and bad. Which do you want first?"

"Just tell me," I said, more sharply than I actually meant.

"Malan wants us to postpone night obs so we can talk at their upcoming meeting. How does that sound?"

"So what's the good news?"

"That was it. You get to give half the talk," said a wry JB.

"Joel. Why don't you just do it?"

With each talk I gave I expected to be more relaxed for the next one, but I still loathed the idea. Joel was right. I should participate. It was a good idea to give a status report and explain our work to both management and research sections at one time. Although our annual reports went to the section chiefs and the head office in Windhoek, we never knew who read them. Giving a talk in Etosha would let everyone there know what we were doing and, more important, why.

JB knew exactly what he wanted to say. I'd do the introduction. However, my ten minutes' worth took six practice sessions until I felt comfortable. Then, on the day of the talk, there was a dramatic delay. The night before, an informant had slipped a message to the Outjovasandu rangers that a gang from Khorixas was going to attack six rhinos being held in bomas in western Etosha. Rangers and weapons were everywhere.

When it finally came time for our talk, we felt sheepish. Who really cared about research when rhinos might be killed? Thinking on a broad scale about rhino conservation, and considering the immediate problem, our project was trivial. Scientific questions, such as how to preserve genetic diversity, or how much space rhinos need, or how horns function are important for long-term conservation, but seemed unimportant in this emergency. What mattered now was simply keeping animals alive. People could resolve the more esoteric issues later. The mood at the meeting was somber. As all of us talked about rhino conservation, the relevant events were taking place elsewhere. From Kenya to South Africa, scouts, rangers, and guards were trying to protect the world's few remaining black rhinos.

Etosha's veld was deceptively brown. Although the 1993 rains had come late, pockets of water were plentiful. Unfortunately, the water had brought with it an upsurge in disease. Malaria was more common than ever in the villages of northern Namibia. Each year it kills more than a million Africans.

Sonja coughed and sneezed. She definitely didn't have malarial symptoms, high fever and chills, but she had something, although it seemed that, whatever it was, she was getting well. Mainly I worried about Joel. His sneezing fits had been regular, at least one a day for the past 23 days. Trying to stay up at night for obs made him worse. After only four to five hours of sleep each of the last ten nights, Joel was listless. Even the zebras that scurried at the water below the observation tower at Calcrete Basin and the rare eland at the woodland's edge didn't excite him.

Sonja and I watched Wilbur, the solitary male wildebeest. With his long stringy mane, upturned horns, and tinges of gray and silver, he was exactly like all the other male wildebeests, indistinguishable in form or shape. We knew him by his outpost. On sentry duty, he ignored us but was vigilant to all else. At night, his snorts and puffs alerted us to lions and hyenas. During the day, he surveyed his territory.

Using behavior to identify animals rests on the dangerous assumption that individuals are recognizable by their behavior alone. If animals have strong physical similarities as well as similar behaviors, there's no true way of telling one from another without marking them. We designated any habituated male in Wilbur's spot as Wilbur. We thought it could be one individual over the field seasons, the same male always lying on or standing not far from the same spot; the others who visited were always nervous.

Among the odd behaviors of male wildebeests is wallowing. Males rolled on their backs, kicked their legs, and appeared to urinate on themselves while rolling. Such self-annointment may advertise body condition; anyone who has smelled a goat or bison during the breeding season can attest to their rank odor. Territorial males energetically protected little plots of ground, some as small as 200 meters by 200 meters.

Besides the smell of urine, the wind carried grass seeds and pollen. We were miserable. Joel was like the walking dead, breathing heavily just to oxygenate his body. As he sat during his thirtieth morning of inactive stupor, black-winged stilts, plovers, and a pale-chanting goshawk visited. Two warthogs were oblivious to the grandstanding of Wilbur. Five new geckos sunned themselves on the thatched ceiling. Joel used his fourth large handkerchief in 24 hours.

That night, JB dragged himself out for another round of observations. We were on the car with Sonja in bed when five lions sauntered into camp. They knocked over the coffee pot and growled at our plastic trunks. Then they came toward us, but campsite odors tantalized them away. A 150-kilogram black-maned male approached the laptop computer. Even the hard Samsonite case would be no protection from canines honed in the Pleistocene by natural selection. Fortunately, they lost interest, drank, and wandered toward the woods.

At 7 A.M. Joel was already lethargic. Drops of sweat protruded from his brow despite the morning chill of 50°F. Congestion pressurized my sinuses. Our heads ached. We needed a physician's help and soon. It was time to leave.

Joel rested while Sonja drew pictures. I tried pouring petrol into the spare tank, but the spouts no longer fit. A few whacks with a hammer reshaped them just enough. I heaved water jugs and trunks onto the rack. Joel slept, unbothered by the flies I could see crawling up and down his face. Two hours passed.

"Watch out for the tricky curves, rhinos, and elephant holes," he mumbled as I drove.

I wondered what he'd been thinking. In Okaukuejo, we dropped Sonja with her surrogate family, the Versfelds. We didn't know how long we'd be away.

"Spleen—okay. Liver—feels normal. Kidneys—okay," announced the physician in the Outjo hospital room. "Let's try some blood," he said, but then missed the vein twice.

"Maxillary sinusitis, as bad as I've seen." He tapped a blunt finger near Joel's cheekbone. Joel jerked away in pain. I had a moderate case—eustachian tubes clogged, glands swollen, and a bright red throat. Finally, Joel stood, but nearly fell, grabbing a railing at the last minute.

"I think you should take your pills now and rest," announced the nurse.

An hour later we sat in the Onduri Hotel. From the bar, Billy Ray Cyrus was blaring his American country hit, "Achy Breaky Heart."

*　　*　　*

The preamble to Zimbabwe's 1992 conservation strategy for black rhinos reads: "Zimbabwe is acknowledged to have the largest surviving numbers of black rhino in the world—approximately 2,000 animals." At a time when most of Africa was struggling to maintain just a few pockets of rhinos, the Zimbabwe figures were stunning—100 at Mana Pools, 600 in the Chizarira and Chewore, 200 each in Charara and Chirisa, 150 in Matusadona, and another 200 in the Hwange region. Efforts on communal lands, the sharing of wildlife-generated revenues among rural residents, and the passing of information on illegal hunting had all helped retain the rhino stronghold.

Zimbabwe also employed aggressive wildlife management. Their philosophy was simple—wildlife is a resource to use. With explosive human population growth and continuing demands for land, Zimbabwe's parks and wildlife must support themselves. Utilization is controversial, but the successes illustrated the logic outlined in the conservation document. "When crocodiles were endangered 20 years ago, Zimbabwe embarked on a vigorous programme of crocodile farming which not only restored the species to abundance but also resulted in a sustainable multimillion dollar industry." That was written in 1992.

An ingredient of success is flexibility, an ability to change as the times demand. A tactic that worked in the past may not serve so well at another time or when circumstances differ. By 1993, Zimbabwe's rhinos were under increasing attack by Zambian poachers. Two of Africa's most talented wildlife veterinarians, Michael Kock and Mark Atkinson, had been busy dehorning every wild rhino left in the country. Even if the rhinos were all killed, at least the government and not the poachers would get the valuable horns.

The Wildlife Society of Zimbabwe explained the dehorning, publishing their official notices in four languages:

> Because our rhino are being poached and killed for their horns the Zimbabwe Government decided to remove the horns from all the living rhinos.

> All the live rhinos have now been dehorned. The horns have all been cut off the live rhino and taken away to a place of safe keeping by the Zimbabwe government.

> It is useless to kill rhino because they have no horns now. Poachers who hunt rhinos are risking their lives for nothing.

Meanwhile, there were signs that dehorning worked. Although the population of 150 black rhinos at Matusadona dropped to 19, only six white rhinos were killed in Hwange, the site where Janet Rachlow was studying dehorning for her dissertation research. It seemed that the frequency of incursions by Zambian poachers had dropped and the behavioral consequences of dehorning were minimal.

In February 1993 Tom Milliken wrote to the USFWS on behalf of an American wishing to import the horns of a dehorned rhino. Milliken, as the director of the South and Eastern Africa office of TRAFFIC based in Malawi, represents one of the world's most elite conservation NGO. TRAFFIC (Trade Records Analysis of Flora and Fauna in Commerce) deals with the international trade and economics of wildlife products, and it is funded in part by the International Union of Conservation of Nature in Switzerland and the offices of the World Wide Fund for Nature. TRAFFIC falls under the auspices of CITES.

The intent of the Milliken letter was to obtain provisional endorsement of darting safaris, in which the horns of a tranquilized rhino are removed and then imported legally. The crux of the letter was that dehorning was working and that darting safaris should be supported on a provisional basis. Its underlying logic was that funds from management could be plowed back into conservation, much as Zimbabwe's rural programs were already doing.

Joel sent a letter to the Office of Management Authority of the USFWS in Washington, D.C. It started a firestorm. He began: "Let me say first that whether you decide (or have already made a decision) to allow horns from dehorned rhinos to enter the USA is *not* something that I wish to contest. Any decision should, obviously, be based on the most credible scientific evidence available." He pointed out that certain types of evidence were still needed to evaluate dehorning biologically, and that TRAFFIC's advocacy of it in the absence of such information was inappropriate. At the very least, it would be prudent to err on the conservative side and first try to locate adverse effects. In the absence of studies of dehorned black rhinos in Zimbabwe, it wouldn't be possible to know if the effects on them of dehorning were negative or not.

There were many retorts to Joel's letter, but events rendered the issue moot. Estimates of Zimbabwe's black rhinos were revised downward. Instead of 2,000, only 300–500 remained. No one knew why. Poaching was the most credible possibility, but others existed, including drought and starvation. The discrepancies might not have been due to poaching at all. Early estimates may have been too generous, especially given the failure to document substantial numbers of poached rhinos. What became clear was that, of the estimated 100 white rhinos in Hwange, most of which had been dehorned, not more than 10 remained 18 months later. Black rhinos were also under siege, horned and dehorned alike. Tragically, anti-poaching teams were grounded because the government had no funds to pay them. *Newsweek* carried reports of the annihilation. Dehorning in and of itself had not prevented poaching. To stop the slaughter more funds were needed, something that had already been learned from discouraging examples that stretched from Tanzania and Kenya to Mozambique and South Africa.

24

Missing Calves

[Joel]

Dozens of horns—rough and smooth, large and small—filled the dark ministry vault. Some came from dehorning operations, others from animals that had died naturally in the veld. Most were from poached rhinos. "Rundu," "Katima Mulilo," "Oshakati," the names of border towns, were scribbled across the severed surfaces. Confiscated ivory, intricately carved, lay in the shadows.

From inside the chamber, we could almost feel the soul of the rhino. The horns were emblems of death—each pair, sometimes each one, signifying a life lost. Dust covered them. Beginning our chore, Carol grabbed the first six and called out measurements as I wrote. Sonja's drawing book soon lost its fascination for her.

"Dad, why do poachers kill rhinos? Medicine is not in the horns." Her three-and-a-half-year-old mind had captured the essence of the tragedy.

We learned that 20 lions were to be removed from western Etosha—darted and sold to authorities in Bophuthatswana as part of a program to reintroduce them into areas where they'd been shot out.

We remembered the pride living around Zebra Pan and the morning we woke to loud purring broken by deep chuffs.

"Uh oh, lions with small cubs," whispered Carol. We lay in our sleeping bag on the Land Rover, watching nervously as they walked and rubbed against each other some 200 meters away. "Should we climb down?" she said.

"No, it's OK. They know we're here. I'm going to shoot some pictures."

Knowing how rare it was to see such young cubs, I sat up and grabbed the long lens. As an afterthought, I checked that a handgun was within reach.

A lioness grew uneasy. Sixty meters from the vehicle, she stopped and crouched, watching us, tail twitching. Then she charged, ears back, legs extended. Raw power raced toward us, 10 meters of separation disappearing with each second. I held my breath.

I tried to unsnap the holster, wasting a precious second—50, 40, 30 meters left. I cocked the pistol—20 meters, still coming. I squeezed the trigger. Nothing! No shot—still coming. I fired again, deliberately aiming in front of the attacking lioness. A "pow" fractured the quiet.

Dust flew as she braked, her forward momentum carrying her another body length closer. My breathing deepened, and I expelled air in short bursts. Only eight meters away, she crouched and snarled. My finger didn't leave the trigger. Both hands steadied my aim as I pointed the barrel at her chest. She turned away, moving silently back to her cubs.

I looked at Cunningham and, not knowing what to say, muttered, "Pretty exciting, huh?"

Her breathing was also deep. "You're such a jerk! If you hadn't moved to get your stupid camera, she might have ignored us. What if she'd killed us both? What about Soni?"

We could still see the lions, but barely. They had nearly vanished through the bushes. I looked at Carol's widened eyes, just six inches from mine. "Wow," I kept repeating, "Amazing, pretty amazing."

I climbed down the ladder to see if the shots had woken Sonja. Still sleeping, she cuddled with her teddy bear.

Namibia differs from most countries. Its founding legislation calls for the explicit recognition of environmental conservation: "the maintenance of ecosystems, essential ecological processes and biological diversity of Namibia . . . the utilisation of natural resources on a sustainable basis for the benefit of all Namibians, both present and future." Selling lions was one way of fulfilling Namibia's mandate; auctioning black rhinos was another. Translocations to private farms also made sense biologically because it reduced the risk of wholesale catastrophe if poachers hit Etosha hard. As in Zimbabwe and most of the world, if wildlife was to have a place, it had to pay.

Albert Einstein once said that we, as humans, need to become more empathetic "by widening our circle of compassion to embrace all living creatures and the whole of nature and its beauty." His wisdom still applies. But when land or animals bring revenue, people aren't always likely to think of the next generation. Right or wrong, the present is what counts. Land development, profiteering, and utilitarian attitudes prevail.

The idea of transferring 20 lions troubled us. Concepts born in the West—that somehow animals in protected parks were inviolate—didn't apply here. On the other hand, these lions frequently visited areas outside the park where they weren't welcome, and since the 1970s more than one thousand lions had been shot on private lands in Namibia.

African management philosophies vary regionally. In the south, the active manipulation of populations and habitats was common, but in East Africa a laissez-faire approach was more the rule.

We didn't know whether one tactic was better than the next. But more troubling to us than the removal itself were its political implications.

The Etosha predator-prey project developed by Bill Gasaway and Andy Phillips was funded by a huge United States grant, signed and countersigned by Namibian authorities as a joint venture. A large cadre of Namibian cooperators and participants were helping to understand why Etosha's large herbivores were declining in numbers. The project was in its second year. If lions were already being removed before the results were in, then there must not be any serious intention to await those results. Decisionmakers, without consulting Gasaway or Phillips, had either already concluded that predation was an important factor in limiting prey or that the lions of western Etosha didn't matter—ecosystem dynamics weren't a concern when lion populations grew large with marauding lions venturing onto private lands.

In the end, the monolithic project collapsed, not because of distant communications and logistical problems but because the Namibian officials and the foreign researchers were unable to find common ground.

* * *

I left the Okaukuejo store with a special treat, a little doll with an original name, Simba. Sonja would be happy. Jovial tourists sat shaded by umbrellas at the pool. I noticed a familiar face. Hans, a waiter, stood at Okaukuejo's gate, the standard spot for local people seeking rides to towns beyond the park's boundary. Usually outgoing, today he was different, subdued, slow. We exchanged the typical formalities, and then I noticed he was wearing a jacket and tie. I looked more closely. His small round glasses had slid down his nose, and his eyes were serious. The vibrancy in his voice was gone, too.

"Where are you going, my friend?" I asked, gently reaching out to touch his shoulder.

"To Outjo, Mister. My son died from the sickness."

We liked Hans and others from the restaurant. But we didn't often think much about them, their lives away from work. We were too preoccupied by our world to notice.

Depressed, we left for Calcrete Basin and Zebra Pan. The moon was waning. Rhinos would be congregating.

For months we'd had problems with our C mount adaptor, a small device that connected the image intensifier to the weighty 500 millimeter lens of our camera. Its threads had become shredded, and the lens barely fit into a connecting tube.

During a water-hole watch at Calcrete Basin, the adaptor finally gave out. I watched in slow motion as 13,000 dollars' worth of equipment slammed to the ground from the top of the Land Rover. The adaptor itself, of cheap soft metal, was worth $15. Fortunately, the intensifier still worked, but without an adaptor the long lens was useless, and we couldn't observe or identify rhinos. To order another from outside Namibia would take months. The moon was getting higher. I had one choice.

I looked at our assorted tools—screwdrivers, chisels, and hammers. None were delicate enough for surgery on the C mount's cylindrical, machine-cut grooves. Finally, I discovered a tool that might work, the leather punch on my Swiss Army knife. I cleared away two years' worth of flattened metal. With the flashlight hanging from my mouth, I crafted new grooves. Although not as close together as the originals, they were deep.

At first, there was no fit. Then it was too tight. The lens couldn't move. I twisted, searching for a new angle. Metal squeaked. Excited, I turned to Carol.

"Here, try it."

The adaptor held. With her next words, "We've got rhinos out there, at least four," we decided we'd have to celebrate later.

At 3:30 A.M. I pulled out two packaged drinks, leftovers from our last trip to Outjo. "Milkshake" was written on them. When we opened the first box, an awful smell escaped. Both were eight months outdated. Not that desperate to celebrate, we went to sleep.

In the Kaokoveld we searched for gray ghosts in sterile valleys. Dead Kudu Fountain, No Name Spring, and Elephant Camp, places where we had found rhinos, all lay empty.

"Old," said Archie. "The spoor is very old. I don't know where the rhinos are."

Never before had the Springbok River, the Huab, or even the gravel fields seemed so dry. The guinea fowl and the elephants were gone. We hadn't seen a mountain zebra in 10 days. Places with baboons in 1991 and 1992 were also empty. Spotted hyenas still howled at night and in the early morning, but we didn't know what they were eating or why they stayed. Except for three gemsbok and two ostriches, the wildlife was gone.

At !Goma, meaning the "difficult place" in Damara, hands pounded the top of the Land Rover. "Rhino, rhino," screamed Archie, unaware that we too had seen it running across the shadeless plain in the midday heat. Apparently it had heard the Land Rover and bolted from its shady resting spot. We jumped into our routine, grabbing equipment, suntan lotion, and exchanging our flimsy sandals for sturdy boots. Archie and I tracked the 21-centimeter spoor. Carol waited, then navigated the car across boulder fields toward the canyon where we'd disappeared from view. After we'd tracked seven kilometers, an animal with nearly 25 centimeters of regrown horn stood looking at us. She had a top right ear notch and slight tear in the lower left ear.

"It's *Mystique*," I whispered. "She had a subadult daughter named *Mystery* with her during the dehorning in 1989."

Then I remembered how well we knew this cow. Too well. My heart added a few beats.

"Archie, do you remember the cow that charged us in the Springbok River in 1991? She was lying next to a euphorb, and I tossed a rock to get her to stand. This is the one!"

Archie's eyes narrowed, revealing small wrinkles in his dark skin. His blue Nike hat tilted. "Yes," he grinned and extended an upturned hand. I smacked it, and he slapped mine. We'd added another rhino, another bonus, another datum.

We photographed her and I recorded the location with our hand-held Global Positioning System, a gift from Carol's dad. At 7 P.M. we rendezvoused with Cunningham, knowing it was still another 60 kilometers to camp. This was one rhino we wouldn't be seeing for a while.

The next morning fresh rhino spoor was at the fountain closest to our makeshift camp. After four kilometers the prints veered off from the two-track and into a rocky canyon. We began our foot pursuit. By 11 A.M. we saw the rhino, a female. She had a top right ear notch and a slight tear, lower left. Her spoor measured 21 centimeters. The similarity between yesterday's rhino and today's was uncanny. We couldn't have two rhinos that looked so much alike.

"Archie, I think this is *Mystique,* the one from yesterday."

"Get a good picture so we can see, Joel."

Weeks later we were all amazed to see that our guess was true. The photo proved that the rhino was *Mystique*. She had traveled more than 40 kilometers in less than 15 hours! None of us ever suspected that rhinos moved that far unprovoked.

On the same trip we found another female, *Christmas*. She had a protruding udder but no calf. We were puzzled. Archie had seen her near dusk the day before, also without a calf. "Come to think of it," said Carol, "we haven't seen any calf prints."

This was startling because we were in the same study area where Pierre du Preez, the ministry's lead rhino biologist who we had helped with our photos the year before, had confirmed sightings of young calves belonging to different mothers. He and I had checked photos a year earlier, and the sightings were definitely not of the same calf: There had been two of them. Now we couldn't find any calf prints, much less an actual calf, although adults were still here. To confuse matters worse, *Christmas* wasn't one of the animals who had calved last year. All we could figure was that she must have just had a calf recently and then lost it.

If we were correct, the implications of our discovery were important. At desert areas with dehorned rhinos, such as our escarpment site or Doros Crater where hyenas were absent, all calves had survived to at least a year of age. The same was true for horned mothers at desert sites where hyenas and occasional lions lived; all four calves had survived.

The small Springbok River subpopulation held extraordinary significance. Here there were hyenas, calves, and dehorned mothers. But now we couldn't find any calves. If the calves were indeed missing, we needed confirmation from others who monitored the area. Earlier I had written to two of the ministry's Nature Conservation officers who had been involved with rhinos, asking them if they had any knowledge of the missing animals. Neither had responded. So we decided to drive to Garth and Margie's at Wereldsend. Garth might know about the calves—the Springbok River was in his backyard.

Garth and Margie had just returned from San Francisco, where they'd received the prestigious Goldman Environmental Award. Usually, one outstanding conservationist from each continent received the tribute. In 1993, Garth and Margie split the African award. Their enthusiasm and commitment were boundless, but their lives were seriously complicated. This year the ministry would allow the local communities to shoot excess animals. Garth, along with others, was helping the local Damaras and Hereros develop strategies for the harvest of springbok, zebras and gemsbok.

We suspected that they had had little time to visit the Springbok River lately, but the temptation to talk about the rhinos was overwhelming. I couldn't resist. Garth's files were impeccable, and they might shed some light on the mysterious disappearance of the calves. For each animal there were now updated drawings of ear flicks and horn shapes, and measures of foot spoor, matched with dates and locations. It was easy to talk with Garth and Margie about rhinos. There was no deceit or paranoia, only openness and communication.

Carol held Sonja while I inspected the records and drawings. Unsure of what I was looking for, I gazed at pages of records, oblivious to the evidence before me. Finally, I reached a page with a drawing of three notches in the left ear, one in the right, and a November 1992 date. It was an adult cow. An *adult cow*, alone! No calf accompanied it. It was *Tammy*, the animal that Pierre du Preez had seen with a new calf in May 1992. *Tammy's* calf had disappeared between June and November. Ordinarily calves remain with their mothers for at least two years. The missing calf had been no more than eight months old and was still dependent on mother's milk and protection. The calf's absence was both amazing and depressing. Apparently, not a single new calf in this river system had survived since the 1989 dehorning operation, a full four years earlier. The small Springbok subgroup of eight rhinos was now down to six. Two had been transplanted, and now *two*, most probably *three*, calves had died. Biologically, there had been no recruitment, no new calves surviving in this subpopulation, whereas our northern desert population of horned rhinos had increased by 50 percent, from 8 to 12.

Why the calves had disappeared was unclear. The drought was intense, but whether it was worse in Springbok River than at Doros Crater or elsewhere was uncertain. What differed most among the study areas was that mothers were horned or hornless and dangerous predators either occurred or they didn't during our study period. Two calves previously seen with dehorned mothers were gone, and a dehorned adult had an enlarged udder but no calf with her during observations of her on several different days. Furthermore, the missing calves were not in adjacent areas, where we spent time searching. They were in areas where we heard spotted hyenas and saw their tracks often. If hyenas were preying on calves, it could mean that hornless mothers couldn't defend their offspring.

We left Wereldsend and drove south to Doros Crater to recheck on the survival of young animals in areas without large predators. We couldn't stop asking each other if hyenas might have been responsible for the missing youngsters. Of course, because we didn't have carcasses, we wouldn't know for certain. In arguing for dehorning as a tactic Save the Rhino Trust, under Blythe's direction, had published papers saying that hyena densities were low and not problematic. Although we had always wondered whether predation was a possibility, our concerns were heightened. We already knew that throughout Africa there was a strong relationship between the proportion of maimed calves and the density of spotted hyenas. The Springbok River mothers had been dehorned in 1989. If hyenas were responsible, perhaps they had switched to calves because other prey had migrated during this harsh drought.

We wondered how Malan, and Blythe at Save The Rhino Trust, would react to such an alarming possibility. Our sample sizes were terribly small. Should we tell them, now or wait for additional years of data? In the end, we decided the ministry should be told of these data, and soon, preliminary though they were.

We passed the red sandstone at Twyfelfontein, where pictographs of rhinos adorned the stately rock. Doros Crater, our nemesis, lay ahead.

25

The Witch Doctor's Revenge

[Carol]

The familiar sound of rock against rubber, shrub against chassis, returned as we left the gravel road. The red sandstone at Twyfelfontein was behind us. Soon mud and stick huts appeared near the site where, three years earlier, several men had stoned a rhino calf and eaten some of its meat.

As we passed the huts, two young boys, both less than 18 months old, sat crying in a swarm of flies. One coughed deeply. Two girls, both younger than Sonja, watched over them, mucus dripping from their noses. This was life at the edge. Without speaking, we drove on.

Sergio Leone's music muted the Land Rover's drone. With an hour to go before nightfall, distant tabletops and buttes were changing from blue to pink as Doros Crater broke into view. A feeling of familiarity flooded us, memories both good and bad—the dehorning operation in 1991, tense encounters with unknown men and vehicles, the surprise flood in the Guantagab River bottom, the clutchless hellride, and the night Joel and Archie got lost and buried themselves in sand to keep warm. Doros was also where Archie had first told us about his role in the poaching, the witch doctor who had given him emetics to purge his drinking, and the curse laid on him and those who help him.

Omundu onganga, James Chapman had called the Damara witch doctors in 1868. One hundred and twenty five years later, sorcery and magic still thrive in Namibia and elsewhere in Africa. In Kenya's Kisii and Nyamira districts, 44 men and women accused of witchcraft were burned to death during a 12-month reign of terror that stretched from 1992 into 1993. In Namibia's Kavango region, the discovery of a mutilated body missing genitals, teeth, hair, and eyes was a grim reminder of macabre practices and of renewed local fears of spells. In 1991, when we first knew Archie, Garth told us not to take lightly the spells cast on Archie by the men he was testifying against in his poaching case. We had listened but hadn't thought much more about it.

"Burning, burning!" were Archie's words, after the cassette stopped abruptly.

"Damn tape deck" was my brief thought. Simultaneously, the dial on the temperature gauge surged beyond the red zone. Smoke began pouring into the car from the front vents.

"Kill the engine," I yelled. Joel held his hand up, wordlessly answering me. The keys already dangled from his fingers. The Land Rover refused to turn off. Idling, its chugging continued. Caustic odors wafted through the vents. Archie jumped out, grabbing a blanket. Joel followed, and grimaced as he and Archie tried to spring open the hot metal hood.

"Get Sonja out of here, *now!*" he yelled.

I grabbed her and ran, thinking about an explosion, knowing that Joel and Archie were still somewhere behind me. It was as if I were in a dream, in slow motion, running but not gaining ground. I was aware of Sonja, nothing else. My legs kept moving up and down, over the rock debris. At 100 meters I put her on the ground, firmly telling her to stay. Then I looked back at the burning vehicle, gauging the distance between it and us. I needed more room. *This is my little girl, my only child.* I carried her another 20 meters. Sonja sat crying, aware that something scary was happening. Plumes of dark soot billowed from the engine.

Archie struggled with the fire. Already Joel was 20–30 meters away with the new red tub filled with scientific instruments, precious data, cameras and film. Dropping the tub, he raced back, shouting, "Archie, get away from the car."

I ran to the back of the Land Rover, and seized more gear, sleeping bags, a backpack that may have been empty, I wasn't sure. My heart pounded with fear. Then I thought of something. Eighty liters of petrol sat on the roof rack, like bombs awaiting detonation above the burning engine. If the Land Rover blew up, Sonja will have only one parent, maybe none. I climbed up, yelling, "Help me, help me get the petrol."

Smoke from melting plastic billowed into Joel's and Archie's eyes. The blanket wasn't suffocating the fire; it was spreading. We could see petrol dripping from the undercarriage. Joel began emptying our water containers onto flames that fluctuated between orange and blue. I struggled to free the tightly strapped jerry cans, slamming each one down to Archie. Soni sat alone, temporarily abandoned on a distant rock. Deep within the engine, flames would die in one place and then suddenly spurt up again from another. Inside the car the floorboard and carpet were melting. Hoses and plastic fueled the fire. Archie grabbed a shovel and frantically tossed dirt into the engine. Boxes of food and drinks, tools, and trunks littered the ground, radiating out from the burning car. Binoculars, books, tripods, tents, and clothes lay between the Land Rover and Sonja.

After 15 minutes an explosion no longer seemed imminent, even though a small fire still burned. One jug of priceless water remained.

"Does it make any difference if we use the last water?" asked Joel. Too crazed to think, I just shrugged. Khorixas was 120 kilometers away. Joel poured it on the flames, but to no avail.

We had no more water.

After 20 minutes more of throwing sand, the fire was out. Metal and copper lay on the ground, the parts that couldn't burn. All the plastic and wiring had disappeared. The blackened engine and fenders and melted vents and floorboard testified to the problem that had bothered us for months. Our inability to solve it might have caused the fire. Our trips to see mechanics in Windhoek, Swakopmund, and Outjo had been fruitless. They had all said that there wasn't anything wrong. But it was our fault that we hadn't carried a fire extinguisher.

Our hearts sank. The 1993 field season had come to an abrupt, premature end. Petrol still dripped slowly onto the ground. With the filmmakers gone, we were alone. And because no one was expecting us, there would be no rescue.

Sonja was as depressed as we were. "Mommy, I have no clay. Where's my clay?"

As we surveyed the mess, we tried to stay positive. We were all alive. There had been no explosion. Even the injuries were minor: some fumes inhaled, singed hands and hair, and scraped skin. No data were lost. Only the front third of the car had been damaged. But Doros Crater had gotten us. The witch doctor's revenge was powerful.

Airplanes were virtually nonexistent, so the chances of someone seeing a distress signal were negligible. One of our biggest fears had come true. We were stranded. But we had saved all the food. Beer was a nonnecessity and could be used for washing dishes. Salvaged juices and cokes were now priceless liquid.

"Let's have something to drink," said Joel calmly. He was right, what was done was done. No need to harp on it. We were all thirsty, and though the refrigerator no longer worked, the juices inside were still cold.

We needed to make a big decision: Who should go for help? Khorixas was three days' walk, maybe more. Sonja would have to stay. I could wait too, but if men from Guantagab happened by, I might be in trouble. Deep in the wilderness I never worried, but here at Doros Crater we had encountered some rough individuals. A woman and a baby by themselves, not to mention surrounded by food and expensive equipment, might be just too much to pass by. I didn't like the idea of waiting.

It was also risky if Archie walked out alone. He thought maybe no one would believe his story and no one would come to help us. Some whites still distrusted black Namibians. Perhaps Archie and I should go. Joel could protect the gear and stay with Sonja while we organized a rescue.

"Just imagine a white woman and black man showing up at the Khorixas Rest Camp after a 120-kilometer trek across the desert. Think of the owner and all the tourists. Maybe we should do it just to see their faces." Archie smiled, agreeing with me.

"I'd love to be there just to see it. Take a camera," chimed Joel. Then common sense took hold. I thought of my feet and the inevitable blisters. Maybe Joel should walk. He and Archie tracked often, and their feet were tough.

Throughout the rest of the afternoon and part of the evening we discussed options. The thought of going troubled me as much as staying did. Guantagab was just too close.

Before first light, JB made the decision. He pushed his warm body away from mine and whispered that he'd go. Soon he and Archie disappeared down an old zebra trail carrying essentials—flashlight, first-aid kit, biltong, crackers, juice, and sleeping bag.

The next sign of human life came 33 kilometers later: a donkey next to a cart. For 40 rand ($13) the owner was willing to offer Joel and Archie a 20-kilometer ride. Then, as they considered the deal, their luck improved. A fume-spewing, diesel-powered dump trunk, overflowing with mopane wood, lumbered into view. Ten minutes later, Joel and Archie piled into the back with four Damaras. Wind whipped their faces. Wood slivers and thorns stabbed their buttocks. The creaking deathtrap bounced and swayed as it dodged cattle and baboons. By dark they had reached Khorixas.

Meanwhile, Sonja and I sat at the car. It wasn't the sun or temperature that bothered me. It was the uncertainty of knowing what was next. I didn't know whether we should try to rent a car to continue with night "obs" at Etosha, or just go back to the States. I knew Joel and Archie would return. The question was when.

Our food and liquid would last at least two weeks. After breakfast, I scrubbed the dirty dishes with beer. I could smell hops. With a rock I smashed the empty can and added it to our growing trash heap. Before Joel left, he had suggested that I hide the data books and camera equipment in an arranged spot more than 100 meters from the car. If people robbed us, they wouldn't get our most valuable goods. I liked the idea.

At midday I heard humming in the distance. Joel and Archie hadn't been gone long enough, less then 30 hours, and at first I thought a plane was approaching. I was wrong. Over the hill roared a dilapidated white bakkie. A hand poked from the passenger window, a white hand. I was squinting into the sun and couldn't recognize the people. All I knew was that there were two. *It couldn't be Joel and Archie already.*

But when they pulled up, it was. This was a "tow-away" car, Khorixas's finest, a 1960 Ford with no swing arm and no hydraulic power, simply a chain attached to an iron pipe. I couldn't believe this pickup was going to pull us through canyons and sand, all the way to town some 120 kilometers away where Archie was waiting.

The driver, a small stocky Damara, surveyed our predicament. His eyes fastened on our packs, then the food. Finally a smile crossed his face. He had seen the beer. He wanted one now. Joel turned to me, and eyes crinkling, whispered, "Negative."

He then told the driver no, politely at first, then more emphatically. Finally he told the man he could have a few once we reached Khorixas safely.

Two hours later we were mired—the white Ford, the Land Rover—thoroughly stuck in the deep sand. We traded off shoveling, first the driver, then Joel, then I. The more we dug, the more hopeless it seemed.

The sun climbed higher. Joel disappeared up the adjacent slope. When I next saw him up on a cliff, he was taking pictures. *Swell.* I couldn't believe his priorities.

Finally, escaping with some helpful pushing by some Damara passersby, we made it to Khorixas. The driver got his beers and money, and we spent the night at the rest camp with Archie. Dinner was kudu steak, and we ordered ice cream with chocolate sauce for dessert. Soni bobbed her head outside a cage, playing hide-and-seek with a captive bushbaby, a big-eyed, delicate-looking primate.

The next morning we arranged for the Land Rover to be hauled to Windhoek, knowing repairs would take at least six months.

26

The Pelvis and the Lion

[Joel]

By the time we arrived in Etosha, almost everyone knew about the disaster at Doros Crater. Malan offered to rent us his short-wheel-base Land Rover, and we gladly accepted. The "Malan-mobile" would make it possible for us to complete night observations in Etosha before returning to the States.

Malan was preparing to leave for an international donors' conference in Nairobi, where many of the world's NGOs and some government organizations would decide which conservation projects were worthy of support. This meeting focused on rhinos and elephants. As Namibia's spokesperson, Malan hoped for big bucks—dollars for patrols, fencing, and boreholes, and other conservation uses.

Pauline, Malan's wife, helped us hoist a rack onto their tan Land Rover. Then Wilfred Versfeld scavenged spare parts while I rigged electrical connections and an extra battery for our infrared and night lighting systems. We were ready for work.

Despite the availability of this vehicle, we wanted to use it sparingly. It was an older vehicle with a damaged chassis, and the leafsprings seemed to be held together by strips of rubber. We didn't need another calamity.

Before we could leave for water-hole watches, Carol had a request to fulfill at the Institute. A roll of negative film needed to be developed for the new leader of Etosha's Anti-Poaching Unit, Piet Pelser. He'd driven more than 1,000 kilometers to photograph a poached rhino near Owamboland in the north for an impending court case; these critical negatives were prime evidence for the prosecution.

The film differed from the black-and-white Ilford that Carol normally used, and she was concerned about developing it properly. She explained what she was doing, step by step, because Piet wanted to learn darkroom techniques. After carefully following the instructions that came with the film, Carol pulled the negatives from the developing tank. Horrified, she blurted, "Piet, there's nothing here!"

Frame after frame was blank. In shock, Piet raced out of the room. Later, they talked through the procedures to discover what had gone wrong. When he had rewound the film, there had been no tension. The camera was a new, fully automatic model, a type he had never used. He'd just assumed everything was in place. But it was obvious the film had never been advanced. Not a single frame had been exposed. There were no photos to develop.

Even though it wasn't her fault, Cunningham felt terrible. A full investigation of the killing had been made, and this case was especially important because an example was to be made of the poachers. At least the local staff appreciated her efforts.

While Carol performed more darkroom duties, I searched for rhino skulls, hoping to enlarge our sample size of known males and females. Nearly 50 skulls, many poached and with panga slashes across their faces, were squeezed into a room outside the Institute. Equally important, Pauline Lindeque lent me an old file drawer containing records of rhino mortalities extending back into the 1970s. I found that almost half the rhinos had died before attaining full dentition, three premolars and three king-sized molars. Because their bodies were nearly the size of adults, without looking at their teeth no one would have guessed that they had died young, in essence, as teenagers.

But why? Why would so many youngsters have died? Fighting was one possibility. Indeed, the majority of Etosha's natural deaths since 1990 had been due to fighting. Predation was another explanation, but evidence of subadults or adults succumbing to predators is rare at Etosha, as it is elsewhere. Other possibilities for the high death rates included disease, starvation, or wounding by bullets—unfortunately, these couldn't be evaluated post hoc.

I also wondered if one sex was more likely to die than the other. We already knew that males were poached more than females partly because males were less likely to run away. But from the skulls, I was reasonably convinced there wasn't a good way to tell male from female. The bodies discovered in the veld had usually lost their genitals and entrails to scavengers and/or predators, so identifying gender seemed impossible. Unlike with buffaloes or giraffes, whose males grow much larger than females, in rhinos both are about the same size.

We thought about our studies of wild horses. Like rhinos, the males and females were similar in size, but the muscles were attached differently in the females' pelvises than the males'. If this was true of rhinos, we should be able to examine pelvises and confirm which were male or female.

Because the moon wasn't yet bright enough for night work, I spent time at the

Institute asking the management and research staff for pelvises. They loved the idea, because, if I was right, it would be possible to learn more about the sexes of animals dying in the veld. Unfortunately, few recalled the locations of dead rhinos, much less the genders. Still, over the next few days I gathered pelvises from the field, from storerooms, and from the Institute. Some came from known females, but I couldn't find any known males until Kallie Venzke, the large and bearded head of management, remembered one at a tree island on the Okaukuejo Plains some 30 kilometers to the northwest.

We knew the spot, having seen a cheetah nearby the year before. Kallie was certain the pelvis was that of a male, but he didn't know if it would still be there. A few years had passed.

One evening after our jog on the runway, Sonja, Carol, and I drove to the place. Crisscrossing the area, we scared up wildebeest and a secretary bird. Ostriches watched us from a safe distance. Finally I saw several large bones. As we drove closer, two tawny objects bolted from thick brush. Lions. One clawed its way up a tree; the other trotted away, eyeing us over her shoulder. The bones were large enough to be a rhino's, but the pelvis was gone. Carol looked around. Ten meters from the base of the tree, and the remaining lion, lay the sun-bleached pelvis. We inched closer. The lion became more agitated, showing us long, white canines. Then he scrambled even higher. I opened the door and quickly pulled the pelvis into the car, hearing, "Why do you want the pelvis, Daaad?"

Back at the Institute, the new veterinarian, a talented South African named Nad Brain, identified the pelvis as male although he had never before looked at rhino pelvises. He could see that males have larger ischial tubercles than females do. Our successful little exercise resulted in the prominent display of a male and female pelvis at the Institute's entrance. Now ministry employees could identify the sexes of dead rhinos.

Just before we left for our final round of night "obs," our hopes for effective conservation dropped to a new low. A headline in the Windhoek paper reported a 400,000-rand grant from the Development Cooperation Office of Sweden to assess "a largely untapped and underdeveloped water resource." It was yet another attempt to suck subterranean moisture from Namibia's spectacular dry river ecosystems.

Many people, especially those in development agencies, understand the importance of water in arid regions, but almost always the emphasis is on developing resources for humans. Canadian zoologists Tony Sinclair and John Fryxell have suggested that basic knowledge of the migratory patterns of mammals would reveal the folly of interference. Where migration by species such as wildebeest or white-eared kob has been the rule, the propensity for overgrazing is usually reduced. But when water resources become available, both herdsmen and their stock are able to become sedentary and populations increase. When the food supply is over-harvested, the animal populations are the first to crash. The Sahel disaster—where boreholes were drilled, people and stock settled, areas denuded, and famine flourished—exemplified this pattern. So, despite noble humanitarian intentions, water tapping has been ecologically disastrous for large swaths of the African veld. Fortunately, some development agencies are now learning to use ecological knowledge to enhance their aid strategies.

In the arid Namib, usurping subsurface water, even miles upstream, could spell death for rhinos and decrease the diversity of other organisms. At sites below a borehole, vegetation is likely to die, reducing food for browsers, altering cover from predators, and changing the thermal environment for all those dependent on shade. Already the home ranges of the Kaokoveld rhinos are larger than those elsewhere on the continent: *Tammy*'s home range was 1,770 square kilometers, *Pinnochio*'s more than 2,200, *Grogg*'s a mere 1,300. Compared to the areas used by other African black rhinos, those in the Namib are, on average, about 10 times larger. The loss of even one water hole could enlarge home ranges further, forcing rhinos into areas with less food or with more humans. The same might apply for migratory herbivores, and virtually all water-dependent species.

There can be no better example of life at the edge than the seminal role of water in arid environments. But, ultimately, ill-conceived development schemes are likely to render unsullied dry country into desert nightmares like Las Vegas or Los Angeles half a world away.

* * *

Night 12 at the Etosha water holes was happily like many others—a plethora of rhinos, aggressive interactions, and successful identifications—but the day began poorly. In the morning the east winds picked up, torrents that had pestered us for the last month and covered us with dust and grime. Carol optimistically had proclaimed that this day would be different, calmer. But by midday, gusts were hitting 60 kilometers per hour. Zebras disappeared under a veil of grit at 100 meters. Wilbur, the ever-present wildebeest bull at Calcrete Basin, vanished although he lay only 35 meters away. We huddled with our lunch behind the Land Rover watching sand hurtle past. I looked at Cunningham. She fired back a lethal gaze, warning me not to mention her forecast. Sonja gobbled up peanut butter. Without a word, she yanked her tangled blonde locks from her face and tugged her jacket closer to her stomach and chest. Unlike her parents made irritable by the wind, much of Soni's life was outdoors. Changeable weather and animals were a natural part of it.

Just after sunset, the moon illuminated a gray shape. A shy old friend, *Chama*, had arrived at a water hole not on his usual circuit. The front horn was now more weathered than it had been in 1991. *Chama* noticed a large spotted shape slink into view and dropped his head. Nearby a leopard crouched motionless. When two other rhinos appeared at the water, *Chama* didn't move. Neither did we. Thirty anxious seconds passed. With a flick of its tail, the leopard turned and glided effortlessly away. Now I could approach the pool to photograph our latest night visitors, already drinking at the water's edge.

Because I was armed with the newest high-tech night photography gimmicks, these rhinos couldn't escape my lust for crisp data photos. I had an autofocus speedlight with hot-shoe contacts, an external power source, LCD panels, infrared light sensors, and a matrix-matched metering system. My eyes had adjusted to the dark, and I ignored *Chama* to concentrate on the other two rhinos. The 1,000-kilogram hulks stood only 27 meters away. I clicked on the LCD panel, and was immediately blinded by the burst of green light.

Heavy pounding erupted. I still couldn't see. With a sick feeling, I disobeyed all the warnings—*don't run*—and sprinted in terror for the distant Land Rover. The thumping continued, but softer, moving away.

When I reached the front fender Carol looked down at me, laughing. Laughing. Then she turned hysterical, howling. *Cold bitch.*

"I didn't know your eyes could get so big. Now you know what it's like to walk around scared shitless all the time." The snickering continued. "Those were different rhinos. They weren't coming at you. When they got a whiff of me, they ran the other direction. They were nowhere near you."

"If I knew that, I wouldn't have run to the car, now, would I?" My ego bruised, I muttered expletives and walked back to my unphased subjects.

Later, a more sympathetic Cunningham explained that there was really little she could have done. If she had alerted me to the approaching rhinos, still 100 meters away, she would have disturbed the ones I was photographing. And because the new duo had run away from me, there wasn't any real danger. Truly sympathetic now, she added, "Sorry. I would have used the spotlight if they'd been closer."

I knew it was payback time. Her tracking in the desert, the times she'd been treed, and harassed by animals had shaken her. She both loved and hated Africa. Plus our record 19 straight nights of "obs" hadn't been without cost.

But our relationship was good. Being in Africa brought us closer. Working together, day and night, was an experience unlike any other. We relied on each other for safety, for stimulation, for fun. We bounced ideas off one another. We kicked balls for Sonja. We were a happy family. We played, and lived, and loved. Our lives were rich.

America was a distant dream. Amenities were enjoyable but not essential. In Namibia we lived simply, pleasurably mixing challenges, books, and biology. There was no television or video, no news of the latest horrors—Russia, the Middle East, urban America. Namibia's changing landscapes and wildlife offered diversity. The warmth and stimulation of friends more than compensated for our transient lives.

We decided to pursue additional funding so that we could extend our project for another three years. Although studying rhinos was not easy, we had persevered. In Etosha we now knew rhinos in three subpopulations where mothers had horns. Our data demonstrated that only one of 10 newborn Etosha calves perished, despite lion and hyena densities an order of magnitude larger than those in the desert. We also knew that long-term drought in the desert had forced most ungulates to move from their usual haunts to the escarpment. In one area the calves of mothers dehorned in 1989 had disappeared, and we speculated that hyenas might have preyed on them, having little other available prey. Our study had momentum and was maturing. We wanted more data to help us understand the system.

* * *

One day I received an urgent message to call one of my granting agencies in the States. Because the phones at Etosha were down, I'd have to hitch a ride to Outjo, 120 kilometers away.

I stood at the Okaukuejo gate for a few hours, and was delighted to see Malan and Pauline's smiling faces as they drove up and stopped. Then I was dumbfounded.

"Joel, I can't take you. This is a government car. Sorry!" he said, starting to drive away. Pauline looked shocked, and I heard her utter, "But, Malan, it's Joel." I'd thought I was "official."

Fortunately, another government employee stopped. My relief quickly turned to fear as his speedometer climbed from 120 to 160 kilometers per hour.

Later, I met with Malan in his office. Now home empty-handed from the donors' conference in Nairobi he ranted about distorted facts and self-serving politics. At the conference, American and Canadian animal rights groups had attacked dehorning, declaring it ineffective and dangerous. They'd used the letter I'd written to the US Fish and Wildlife Service as primary evidence.

"Others in southern Africa don't know you, Joel," he added, "and they don't trust you." Although Malan knew that we were trying to infuse some science into rhino management, my letter had had devastating repercussions, at least from the southern African perspective.

"Malan, have you read the letter? Did you see what it said? It emphasized the role of science. Besides, it was full of qualifiers."

It didn't matter. The meeting hadn't been attended by scientists. Most participants had a political agenda. "Only parts of your letter were used," he vented, "the parts that supported various anti-dehorning propaganda."

The letter, as well as comments made in the scientific correspondence I had published in *Nature,* had been deliberately misquoted, their significance distorted, and we were being damaged. Understandably furious, Malan asked that we never again use Etosha Ecological Institute as a mailing address.

Two weeks passed. We said our goodbyes to the Institute's management and research staff before departing for Windhoek and the States. Carol printed duplicate photos of study animals to update the park's records. I provided our latest demographic estimates and donated radio collars for the ministry's monitoring programs. Then I hand-delivered my annual progress report to Malan, the same one that we would give to the ministry's director in Windhoek.

It encapsulated what we'd learned so far during our three years. As requested by the government, it discussed rhino mortality, dehorning, and potential population viability. Also, because we'd been asked by the Namibian Wildlife Society to give a public lecture in Windhoek, we wanted to be sure this was all right with the ministry and whether we should avoid any topics for security reasons.

"Namibia believes in an open policy and information shouldn't be concealed," Malan said.

As we packed for the trip to Windhoek, Malan stopped by the caravan. He'd read our report and thanked us for the information. With him was Rudi Loutit, who was shocked to learn of the missing calves.

"Doesn't Blythe know about this?" I asked somewhat surprised, given Save The Rhino Trust's monitoring program.

Rudi wasn't sure. We sipped tea, talking amiably for another hour. Once he was gone, Carol and I recalled past interactions with Blythe—tense encounters about study areas and how data could help shape management decisions—and wondered how they'd affected Rudi's views of us.

At our introduction to the ministry's Rhino Management Group back in April 1991, I had been granted permission to speak, and I had asked about the possible effects of removing female rhinos from the desert population on its long-term viability. Twenty-five percent had been removed and, in other species, the removal of too many females has jeopardized the continued existence of local populations. I hadn't realized then that that question was like tossing a dart at a balloon. Traditional debate, which is healthy for promoting science and conservation, was unwanted here; one didn't question the authority.

Blythe had waved a sheet-filled folder at the group, saying, "We have studied these animals for eight years, Dr. Berger, and have all the sex and age data here. It's fine to take animals out." The discussion was over.

Recognizing that Blythe did know many of the rhinos and not wanting to embarrass her or put her on the spot again, we had sought her advice in less formal settings at least six times, hoping she'd share her insights. Then, seven months later, at a meeting with the permanent secretary, the deputy minister, and Eugene Joubert, we had again asked about the folder and the identities of the dehorned rhinos. This time we had anticipated a stinging indictment. It had come.

"They're [Berger and Cunningham] going to have to work and can't just walk in and have everything handed to them. I'm not interested in science. Conservation must be done!"

We couldn't believe it, and had wondered whether everyone thought we were sightseeing when they didn't see us for months on end. It was as if Blythe seriously believed that we were just hovering like vultures waiting to publish "her" information, and that science had absolutely nothing to do with conservation.

Rudi had generally been so amiable that we believed that, out of deference to his wife, he supported her but was also interested in our findings. As we said goodbyes to Rudi and Malan, Malan indicated that he looked forward to seeing us next year. In a borrowed Volkswagen van with our gear piled to the ceiling, we left.

At the capital, I met with three top officials, including Polla Swart, the director of the Ministry of Wildlife, Conservation, and Tourism, and handed a copy of our report to each. Aware of the demands on their time and the interminable number of reports cast on them, I briefly summarized our results. In response to a request Malan had made for management recommendations, I also suggested that the loss of calves had serious implications. It was difficult to see how to convey the idea that had we not been involved in rhino research, the missing calves would probably never have been reported. I openly wondered about the lapse, whether information had been concealed deliberately or unavailable because of shoddy monitoring. I showed them copies of ministry and Save the Rhino Trust files, comparing copies of our rhino photos with theirs, which included one photo from a reversed negative. At least 50 percent of the ear notches and markings were wrong in the records of the 1989 dehorning. Their jaws fell open. They had known nothing of the problems, nor that spotted hyenas still lived in one of the areas where dehorning had occurred.

We chatted about alternative explanations for the missing calves and the effects of removing females on population viability. Someone asked how the horned population north of the Veterinary Cordon Fence was doing. I explained that it had increased by 17 percent, compared to the population to the south, which had dropped by 25 percent. In response to another question, I said nervously, "I'm not sure what to recommend, other than being careful to make sure your decisions are made on the basis of information and in the interests of rhino themselves, not internal desires to raise money from outside Namibia." They immediately understood my meaning—that Save The Rhino Trust was busy soliciting funds from South Africa, Europe, and elsewhere but their effectiveness in monitoring rhinos was questionable.

Before leaving, I explained that Carol and I felt fortunate to be in Namibia and that our funding was not dependent on our results. We had no financial stake. We simply wanted to learn and to help; the best way to do that was to report our findings objectively. I thanked the officials for their time. Dannie Grobler, head of the ministry's management division, offered as an afterthought that he was pleased that outsiders were studying rhinos. I left for our evening seminar.

Carol began the talk. She explained what we were doing, our scientific goals, and our methods. I followed with our preliminary research results. The turnout was good, although there were only three ministry employees in the audience, and we were disappointed that no black Namibians had come.

Blythe was there. Afterwards, she expressed surprise to hear that spotted hyenas lived in the Springbok River. I was surprised that she didn't know, replying, "You've haven't seen any? What about tracks or dung?" She hadn't, but we understood why. Blythe rarely spent nights in the Springbok study area. By the time she had arrived later in the day, wind and trampling by other animals would make tracks less obvious.

But a comment by Johnnie Roberts, one of her coworkers, stunned us. John, a fundraiser on behalf of Save the Rhino International in England, explained to us that it really didn't matter whether calves lived or died. What was more important was that actions were being taken to save rhinos. Donations for antipoaching patrols or education didn't have the same appeal as dehorning. Horn removal was sexy, a powerful magnet that appeared worthwhile to the public.

In disbelief I stared toward Cunningham. Here was a spokesperson, for Save the Rhino International, telling us that it didn't matter whether dehorning worked or not, because the tactic was good for raising money!

A small blurb, "Be Aware of the Green Clones," appeared in the Endangered Wildlife Trust's magazine, alerting its South African audience to the dangers of official-sounding groups with names like The Roan and Hippo Foundation or the Breeding Endangered Species Trust. I thought of the foibles of people all over the world who haven't spent time in Africa but want to help. Like Lewis Carrol's naive Oysters, lured by the Walrus and the Carpenter with a good story, they'd be unable to evaluate misleading claims of conservation successes.

As we drove to the airport for our flight to the United States, Carol and I talked of many things: of lions—and calves—and pelvises—and how our year had been.

27

Horn Traders

[Carol]

Nevada; September 1993.
We launched into our other life as the university's fall semester began. Joel taught, and we wrote grants, analyzed data, and provided overdue fiscal and progress reports. Days vanished at a frenetic pace. At home, practical things also demanded attention. Pipes split from last winter's freeze, dying trees, and extensions on last year's taxes from the Internal Revenue Service had to be resolved. Did our colleagues seriously believe we were taking one long vacation in Africa?

Just as daunting were our early morning confrontations with an uncivilized four-year-old. Soni couldn't understand why she had to be waked up, dressed, and driven to preschool. I had to work, and she needed to be with kids. Early to bed and early to rise didn't work as well in a house with curtains and electric lights.

Rhinos in far-off places became unreal, shimmering and fading on distant pans as we rejoined the realities of mainstream America. But even on the eastern slopes of the Sierras, we couldn't escape African politics.

The *Wall Street Journal, African Wildlife Update,* and *Science News* called, all anxious for the latest information on rhino dehorning. And the recent news from Namibia was unsettling. Despite our confidential report to the ministry on missing calves, the Windhoek newspaper had heard of our results and dismissed them as rubbish. The government had decided to move forward with dehorning and stockpiling horns. Rhinos in one of Etosha's fenced areas were scheduled for horn removal. The permanent secretary, Hanno Rumpf, knew the monetary value of action-packed projects, and an American media group had agreed to defray helicopter costs. Meanwhile Joel focused on his 8 A.M. lecture.

We heard about Africa vicariously, through the media. Two articles particularly caught our attention.

The Associated Press headlined ghastly news: "Lions Kill Tourist in Etosha National Park." A young German sleeping at the Okaukuejo waterhole had been partially eaten by two lions near the spot where, more than a year earlier, Joel had wondered about his own safety. The old female had approached very near where JB had sat on the bench. Lion tracks had been found on the tourist side of the stone wall the day before the unsuspecting German tourist was eaten. The old lioness and her male consort had been killed.

A headline, this time in the *San Francisco Chronicle*—"Princess Arrested for Possession of 22 Rhino Horns"—was equally distressing. Bhutanese Dikiy Wangchuck, a Cambridge graduate, had been apprehended in Taiwan with horns purchased in Hong Kong worth $740,000. Although rhino horn couldn't be traded legally, it would still be marketed by less scrupulous methods. The question was whether the world should pin its hopes on the prevention of trade or its legalization.

When CITES, the Convention for International Trade in Endangered Species, convened in Kyoto, Japan, early in 1992 the overriding impression in southern Africa was that rational decisionmaking had been shanghaied by western protectionist lobbyists. Trade in ivory and rhino horn, and the harvesting of whales, were some of the emotionally charged issues.

As usual, Africa was split over legalizing trade in elephant and rhino products. Zambia voted one way on some issues, Namibia the opposite. Tanzania might side with the Zambians, and Botswana with their western neighbors.

The Southern African Consortium for Ivory Marketing (SACIM), composed of Zimbabwe, Botswana, Malawi, and Namibia, championed ivory utilization. Much of the rest of the world did not. The consortium argued that views from the United States, Canada, Switzerland, Japan, and other countries that maintained utilization schemes for wildlife within their own borders were pure hypocrisy, because the same right should be allowed in southern Africa. Of course, the harvested species were not listed as endangered.

George Hughes of South Africa's Natal Parks Board went so far as to propose down-listing white rhinos so that their horns could be traded legally. The request had little support from the delegates representing the 104 nations in attendance. (His request has fared better in 1994 and in 1996.)

That countries, financially strapped or not, should earn money for their conservation programs by selling products from their national parks is anathema to many westerners. But to parks in poor countries with large numbers of animals and limited space, sustainable use generates revenue while it protects ecosystems. In Natal's Umfolozi Reserve, 1,136 wildebeests, 1,524 warthogs, 36 waterbucks,

7,468 nyalas, 9,810 impalas, 973 buffalo, and about 50 rhinos were removed over a three-year period 20-some years ago to prevent damage to the 900-square-kilometer park. This, in essence, is wildlife management.

In the 1970s North Yemen had imported some 40 percent of the world's rhino horn for "djambias," dagger handles, many of the finest elegantly studded with gold and silver and worth more than $10,000. Yemeni males over the age of 12 wear these daggers. Those with money prefer rhino djambias. Others use camel hoof, and the poor man's equivalent comes from cattle horns.

During the next decade, the importation of rhino horn into Yemen dropped almost 95 percent, from about 4,000 kilograms per year to a mere 250. Assuming the horns on an average black rhino weigh 2.88 kilos, that means that instead of 1,388 dead black rhinos a year, the imports came from only 87! Such figures are deceiving, of course. Horns from other rhino species surely would have found their way into Yemen, and some rhinos probably died naturally. But considering that black rhino numbers had plunged from some 60,000 in the 1970s to less than 3,000 in 1993, it remained clear that natural mortality was not wiping this species from the earth. A persistent and lucrative illegal market pushed the decline.

Adding to the already voracious appetite for horn, after the 1990 unification of North and South Yemen and the inevitable increase of unpatrolled borders, illegal importation resurrected itself there.

But the use of African rhino products began well before the 1970s. Between 1849 and 1895, more than 11,000 kilograms per year may have come from East Africa. Some 170,000 rhinos may have died.

Today, Yemen's recent thirst for rhino horn is dwarfed by the Asian market, which has existed for centuries. Cups and decorative bowls have been dated from the years 600–900 and 1650–1900. Elegant Taoist carvings with ornate figures were in Europe by the nineteenth century. Collections are displayed at Ireland's Beatty Library, the Netherlands's Museum voor Land en Volkenkunde, and the Fogg Museum in Boston and the Field Museum in Chicago. Unlike times past, most of the horn on the Asian black market is now in the form of medicinal powder. A kilogram fetches from $3,700 to $16,000. The horns of Asian rhinos are worth more.

The problem is considerably more complex than the extinction of a species. "It involves the extermination of a cultural practice that has existed for 2,000 years or longer," argues botanist David Charlet, who worked for years in San Francisco's Chinatown pharmaceutical business. He points out that traditional Chinese medicine incites typical western skepticism about herbal healing, and he underscores the fundamental differences between eastern and western scientific philosophies. Nowhere are these differences more apparent than in the results of a survey about rhino products and their future use. In Taiwan, Kristin Nowell and her colleagues asked 125 traditionally practicing physicians whether rhinos should be conserved; 124 felt rhinos should remain "so that they reproduce and the source is not extinguished."

The dilemma cuts to the heart of human rights and conservation, and polarized views about what is human and what is not. Should westerners push to save species and ecosystems, ignoring foreign sovereignty and the potential loss of long-standing cultural practices? Government-mandated control of deeply rooted beliefs is no less insidious, argue some, than the U. S. government's refusal to allow Lakota Sioux their spiritual ghost dances late in the last century. Is "pharmacological extinction" any different?

"Absolutely," argue conservationists, "when it leads to biological impoverishment."

As 1993 wound down, the United States became increasingly intolerant of the glacial pace at which Taiwan moved to curb the illicit market, despite the arrest of Princess Wangchuck. The National Security Council recommended limited trade sanctions against Taiwanese animal exports—an $8–16-million-a-year business. Although the sum involved was relatively small, Taiwanese politicians were displeased by Washington's interference.

The semester ended in December. Soon we'd be flying to Namibia. Joel had explored the possibilities of continued support and now prepared a lecture for the Wildlife Conservation Society, the new name for New York Zoological Society. This would be an audience filled with real-world conservation biologists, people who knew the cold realities, the bite of politics, and the challenges of foreign bureaucracy.

He returned full of enthusiasm and optimism. Though no one had promised money, there was a lot of interest in the fact that, outside of South Africa, Namibia harbored most of the continent's black rhinos and the largest unfenced and national park populations in the world. The Wildlife Conservation Society wanted to hear more. They'd consider additional funding.

*   *   *

Most of our extended family lived in southern California, and we liked to be there for the winter holidays. We wanted Sonja to know everyone better, and our six months in the United States was short and busy.

This time the 11-hour drive from Nevada was seeming longer than usual because Joel was coming down with the flu, perhaps a bad cold. Aches and chills seized him only a few hours into the drive. Although not completely unsympathetic, I still thought, "Here he goes, macho man down for the count." There was a precedent. All three of us had caught a horrendous flu in Namibia. Soni had vomited and suffered from high fever while I shivered from mine and tried to help her. Meanwhile, Joel staggered to the bedroom, fell diagonally across the bed, and didn't move for the next day and a half. I'd heard of this scenario lots of times. Life goes on for the sick woman, but the man falls over, and life stops as far as he's concerned.

I did feel sorry for him, but there he was, leaning against the car door, dead to the world, and I wondered if we should forget the whole visit and head back to Nevada. He surely wouldn't be seeing anyone. When he woke and moaned that his ears were swelling and that "this was 'it'," implying imminent death, I groaned inwardly. "Joel, we'll buy you some flu capsules and tylenol." I wondered how often people thought their ears were swelling. His certainly looked normal.

To break up the ride, we stayed with friends that night. The pills didn't do much. Sleeping next to him was like being too close to a radiator. The next morning, with only an hour-and-a-half drive to his parent's home and a bed where he could rest, we left early. I drove.

Joel kept mumbling that his head was swelling and waved fingers over his forehead and around the back of his head, forming a crown, trying to show where the pressure was. Neither I nor his parents could see anything unusual, but we agreed he should be checked.

At the emergency room, the nurse listened patiently to his story before responding, "Sir, that would be most unusual." Then she told us that the hospital often received many crank visits from the nearby psychiatric hospital. We returned to his parent's home with stronger pills and hoped he'd feel better by morning.

At first light Joel turned over to say good morning. I thought, "No, it can't be," and reached for my glasses. "Joel, you were right. Your head is swollen. We're going back to Emergency."

From forehead to midcheek, including at least one ear, he was deep pink and inflated like a balloon. At the hospital a round robin of doctors passed through the room. They were fascinated, "Hmmmm, and you just returned from Africa." We told them we'd been back for months but the refrain of Africa remained.

The first doctor discovered a small sore at Joel's hairline, apparently the source of a major infection trapped between his skull and scalp. An infectious disease specialist took over, a Vietnamese man with a sense of humor but difficult accent. They talked of Africa and Asia, tiger bones and bear gall bladders, as others prepared a highly concentrated intravenous antibiotic. Joel could go back to his parents' home if he agreed to return if the fever continued.

The inevitable happened. His head grew larger, and he bore a strong resemblance to a ruddy Mickey Mouse. His head, usually fairly narrow, was a swollen triangle, brows and cheeks extended, above a ridiculously normal chin. I was scared. I'd never seen JB this way. Even when he was sick in Africa, it wasn't this dangerous. I thought of invincible viruses and flesh-eating bacteria that are now killing people because they'd developed resistance to most available antibiotics.

Joel was hospitalized immediately. Through a shunt, powerful antibiotics dripped into his veins. I visited him in the hospital the next day and the day after.

Three days before Christmas was a special occasion. Soni leaned over and kissed Joel, whispering, "Happy birthday, Daddy. How do you feel?" He tried to wink. I was really nervous. Maybe he was right, and this *was* it. His head and ears were still swollen beyond belief. The disease specialist was on call day and night. There was one important improvement. The fever was gone.

At last he left the hospital and recuperated at his parents', returning morning and night to the hospital for antibiotic injections through the shunt. Finally, we drove back to Reno with Namibia on our minds.

IV

Year of the Human
[1994]

28

Rhino Rhetoric

[Joel]

January snow blanketed northern Nevada as we prepared for a fourth field season. Despite sunny blue skies, high temperatures hovered near 20°F. We visited physicians, dentists, and accountants, and each day stacked supplies in our living room. We also bought last-minute items for Namibian friends—gun parts, a camera, a day pack and tent for Archie, a book on elephant lore for Malan. But there was still no one to housesit or care for the dogs. Airlines in the United States, Europe, and South Africa had yet to decide on charges for our excess baggage. Even getting to Los Angeles International Airport would be difficult because of the massive freeway destruction from the devastating earthquake only a week earlier.

One thing kept worrying us—the six-month silence from Namibian officials.

Back in Windhoek, the burnt Land Rover was our primary concern, especially after we remembered an unlucky colleague whose entire field season in Zimbabwe had been grounded while he waited for spare parts. The last we knew, our vehicle had been towed to the capital and, with charred engine and melted tubing, was to sit idle in a guarded parking lot until parts arrived.

Taking a taxi directly to the garage, we looked for our yellow license plate, number N69745W, to confirm that a shining white Land Rover was really ours.

It had been equipped with new electrical harness and other parts imported from England and South Africa. We were ready for business.

Our euphoria was premature. Although the ministry's director had passed our express mail package along to Henti Schraeder, the head of research, this year differed from the prior three. Our research clearance hadn't been processed.

"Dr. Berger, what are you doing here?" said Henti in a startling, calm voice over the phone. "You've already published a paper."

Dazed, I asked if he'd seen my earlier progress reports, indicating that we were returning. "How could our research be over? There's a year to go." I pointed out that the other Namibian officials, including Malan at Etosha, knew that we'd planned to come back. During our exit meeting last July, officials had said they looked forward to seeing us. Henti would have been at the meeting that day, but he'd been ill. He had been briefed.

Impassive, he replied, "Let me check the original proposal and speak to Malan. I'll get back to you."

"Henti, we'd really like this cleared up. Can I stop in and see you?"

"No. I'll get back to you."

Cunningham knew something was very wrong. "Is the study over?" she asked wistfully.

As we waited for the return call, we replayed the year's events, something we'd already done at least a dozen times back in the States. I had written four letters to Malan, including mailing a new book on African leopards, and also several letters to different officials. There had been no response, just silence. Could they really be upset over our publication on horn growth rates? Malan had been a willing coauthor. Maybe it was the progress report. Too honest? We had pointed out the problems in the previous rhino identifications and that calves were missing. We had suggested that a high proportion of Etosha's rhinos were dying before adulthood. The more we talked, the less we understood.

Malan had asked us to give him a report detailing our management concerns. Now we questioned whether doing so had been prudent. We remembered his diatribe about my letter to the USFWS that had discussed why data should be used in decisions about importing rhino horns. We also knew we had upset Save the Rhino Trust with our assertions about missing calves. Intending to help, we had stumbled into the politics of conservation.

Waiting in our hotel room, we lived moment to moment, our spirits rising and falling, as if on a roller coaster. People we tried to reach always seemed to be at lunch, on vacation, away from phones. No one was available. We understood that decisions take time. Permits might take even longer. This was the Africa we knew. Although permits had been waiting for us before when we'd returned from States, our problem had been expecting permits to be ready this time. We tried to stay calm. Without expectation, there would be no disappointment.

If there really was a problem, why hadn't we been notified while we were back in the States? Surely, no responsible person would have allowed us to travel all this way, spending $10,000 on transportation and field supplies. We decided we must be paranoid and dismissed our worries. Besides, Malan had always been a faithful project supporter, renting us his vehicle and defending our research to detractors.

The phone rang. It was Henti.

"You should go to Etosha and see Malan. We'll give you a temporary permit to visit."

So we drove north from Windhoek. The Land Rover was reborn, rocking and tilting in its boatlike way. We played a Jackson Browne tape, remembering old times. Sonja announced with longing, "My eyes will be hearts!" She wore a dreamy look of anticipation, our love-struck four-year-old eager to see her friend, Pieter. Reddened Kalahari soils came into view. A yellow-billed hornbill dipped and fluttered, disappearing deep into the mopane forest.

At Okaukuejo the veld was green, refreshed by recent rains. A raucous tour bus rounded a corner too quickly and sent springbok pronking and wildebeests kicking on ungainly legs. Despite the splendor, we couldn't escape the grip of Henti's words.

In the capital we had learned that the permanent secretary, Hanno Rumpf, had met with Malan, Blythe Loutit, and others just a month before our March arrival. Both Blythe and Malan argued that our research was bad for Namibia and should be discontinued immediately. Rumpf had circulated a confidential memo to other ministry officials indicating that outsiders should no longer have access to ministry information and that all exchanges with foreigners about rhinos and elephants were forbidden unless approved by the permanent secretary. Whether that meant us was unclear.

Later, the impala researchers Aron Rothstein and Wendy Green told us that a ministry biologist had told them, a month before our arrival, "Joel and Carol are no longer welcome in Namibia." When they asked why, no answer was given. The topic faded.

Shocked and confused, we considered our next move—the caravan, beans, and bed for Sonja. We couldn't help but note that no one was talking to us. Tomorrow we'd see Malan and Pauline at their home.

*   *   *

"Some people in the ministry are against you and believe that your motivations are political, that you want to criticize some of the ministry's handling of rhino management at Etosha." Others had suggested that we were anti-dehorning even before coming to Namibia and that we operated with our eyes closed. Had it not been happening to us, the comment would have been laughable. Were we really political saboteurs?

Malan continued. "Your progress report was rejected—it was unacceptable! Peter Erb and I are putting together a full response. Perhaps you could comment. Otherwise, we may try to publish it in an African journal where you'll be unlikely to respond."

I was dumbfounded. My jaw fell open, and I looked at Cunningham. She too stared in disbelief.

"People are questioning why you are here," he said. "Someone compared the situation to a physicist publishing top-level government secrets about a new weapon. Credibility and security are both lost. What you have done is very dangerous and could cost us hundreds of thousands of dollars."

Pauline looked up, horrified. "Malan, this is the first that I've heard of any of this."

Malan intervened swiftly, like a commander in the military, raising his hand, flicking his index finger into the air. Pauline was stilled. Instantly, the room fell silent.

"But if there was a problem, why weren't we notified?" asked Carol seeking clarification. "We're not interested in confrontation. We would have done things differently. We asked for comments on the progress report and didn't hear a thing until just now. Do you think it was unreasonable for us to think everything was fine? Joel wrote, too. Shouldn't the ministry have told us that the progress report had been rejected?"

Malan had no answer.

In the Land Rover, I looked at Cunningham. "I didn't know progress reports could be rejected. Did you?" She shrugged. We returned to our caravan, now knowing a new meaning of the word "low."

A few days later, I met with Malan and showed him a manuscript soon to be published in the American journal *Science*. It contained much of the same language that was in the progress report, and would now be a huge problem. He'd read the report back in July and hadn't been negative, though during the intervening eight months there had been the unprecedented silence.

The manuscript for *Science* dealt with general aspects of rhino conservation and human intervention across Africa. The Namibian emphasis and the specific management recommendations that were in our original report to the ministry had all been purged. Perhaps he'd realized the paper's merits.

Wrong. Malan went ballistic. His face turned beet-red as he uncontrollably slammed his fist onto his desk, shouting "abusive and nonobjective." For a moment I thought he was going to punch me. "There is no one within your corner in the ministry, and my support is now eroded." Malan continued. "We are like a family, we Afrikaners. You have united us all."

I couldn't believe this was happening. He accused us of studying aspects of rhino biology that hadn't been included in our original proposal. He ranted, "You didn't even find the carcasses, so you don't know that the calves died . . . You'll have to write a new project proposal and your request will have to be handled by head office [in Windhoek]." I thought back to our initial 1990 proposal and its goals of assessing pregnancy, calf production, survival, mortality, horns, combat, predators. I didn't understand why Malan thought we were doing anything different, but I agreed to draft a new proposal for 1994. Two days later it was ready, directly modeled from the original.

Unwilling to listen, Malan proclaimed, "It's substantially different."

"How can that be, Malan?" I said, "The goals are the same. The methods are the same. We based this on the first proposal, the one that the ministry, and you, approved."

Despite my objections, the proposal had to be reevaluated.

Then he asked why we were publishing preliminary findings. I pointed out that the American system may not be the best, but we felt obliged not to wait years before releasing results. "There is a tremendous amount of interest, misinformation, and hype about dehorning, little science and too much sentiment. The media is regularly making claims that there are no ill effects. Where do they get this information? Millions of dollars have been invested in rhino conservation."

We didn't think it was right to use dehorning as a banner to attract money if possible negative effects were ignored. At the very least, we felt we should put forth some biological information on the status of dehorning.

The ministry had denounced us, suggesting that they'd be better off not knowing details, that they would never have known about the missing calves if not for us. We were hearing a variation on a lecture Malan had often given Bill Gasaway during the collapse of Bill's project. The outside world did not need to know anything about Namibian conservation practices. Rumpf's internal circular about secrecy and no further communication with outsiders made that more than clear. It was a reversal of what Malan had told us before our 1993 lecture—that Namibia was open and we should communicate details freely.

Our differences of opinion about the public's right to know were distinct, and currently nonnegotiable. Malan disagreed with Carol and me. We parted company.

We were free to stay in Etosha until our new proposal was reviewed, but we could not continue our research. Rhinos could not be studied because waterholes could not be visited at night. All privileges were rescinded. We couldn't go beyond tourist areas and had to abide by their curfew. The ministry would decide whether our study could continue.

When I asked how long a decision might take, there was no answer. The permanent secretary was away on a trip, and Malan himself was leaving soon for overseas. When I pressed him, asking if the reevaluation could take six months, he said, "I don't know. Perhaps."

We debated whether to sit it out or leave. Malan had been very clear that our project lacked internal support. Although the ministry had not literally canceled our research, they had effectively halted it. The project had crashed to a standstill. A long delay, if not outright denial, would control us effectively. Without clearance, we could only sit until our visas expired. Then we'd have to leave.

Something else was disturbing us, but we weren't sure if our qualms had substance. The remark about scientists who reveal state secrets nagged us. Some Namibians were volatile and ferociously committed to dehorning. If they thought we were interfering or had a "political agenda," who knew how far they'd go to remove us.

It was like a bad movie. Going underground and waiting it out in Swakopmund or elsewhere where friends offered us refuge was an option. Others suggested legal recourse. None of the choices were appealing. This wasn't a vacation. Time and money were being squandered, especially money from agencies that had invested in us. We wondered if we'd be better off going back to America and working on data. Doing nothing in Okaukuejo seemed fruitless.

"What about Archie?" asked Cunningham. *Uh oh,* I thought, remembering our deal. I had told him that he shouldn't believe people if they said he was no longer working with us, that news would have to come directly from me.

Archie was already expecting us in Sesfontein where, of course, there wasn't a phone. His post office box was in Khorixas and there was no way to know when he'd receive a letter. We'd have to tell him, even if it meant three or four days of driving. On the way we could visit the Gasaways and the rangers in the western part of Etosha. But Garth and Margie were in Kenya, and Rob Simmons and Phoebe Barnard were in Sweden.

We still had one hope. Malan had invited us for a last meal. But we couldn't shake the feeling that he could help if he wanted to. We finally decided to stop by his house Saturday morning to decline the invitation and suggest a compromise.

We explained to Malan that we had come to work and had just bought new tires, vehicle insurance, and months worth of food. We were fully prepared to gather more data to bolster sample sizes. Our lives were in turmoil, split between widely different responsibilities on two continents. We loved Namibia and its people and believed in their conservation aims. But we weren't ready for more battle. It wasn't the arduousness of fieldwork that we couldn't tolerate. It was the political interference, the backbiting, the egos. As Carol put it, "Were we there only to endorse official dehorning policy?"

We offered the compromise. Perhaps the ministry would offer a temporary permit so that we could work in Etosha while the new proposal was being evaluated. Without some inkling that support might come, it seemed futile for the three of us to sit and wait for permits that might not materialize.

We were too optimistic. Malan found the idea of a temporary permit unacceptable. Nothing had changed. He couldn't assure us that the review would take less than six months.

We weren't surprised. So, despite 197 nights with rhinos, more than 1,030 hours of observations and 100-some known individuals, more than 5,000 kilometers covered during 126 sampling transects, our efforts to study black rhinos came to an end.

* * *

The veld was still fresh and beautiful. We tried to absorb it with our senses. Savannas gave way to trees, then kopjes and hills. The grass melted away. A bateleur drifted overhead on thermals. Mountain zebras and giraffes emerged on barren plains. Finally, goats and then Sesfontein came into view. Painfully, Carol and I asked each other if we were really ready to give this up. America would look pretty tame.

Archie waited. He was prepared for a month with rhinos in the western deserts. We didn't know how to begin to tell him that our clearance to study rhinos was gone, the project was over.

Excited to see us, he gave us a quick tour of the new house he'd built. A tall ceiling, metal roof, and windows separated his home from Sesfontein's other dwellings. The typical assortment of thatching, sticks, and mud was still evident, but Archie's home was different. His was much larger, with several distinct rooms.

"What do others in Sesfontein think about all this, your money and your work?" I asked.

"It is good, Joel. I think others wish they had jobs. I am lucky. I have been asked to be on the Sesfontein council."

This meant that Archie would represent the community in natural resource affairs, serving as an intermediary in discussions with the Namibian government. It was a form of local rule, and Archie would be a key player. The poaching incident had faded to a bad memory. Now he had respect.

It was my turn. I asked Archie to step away from the others and then told him that our project—the joys and despair, the tracking and close escapes, the victory celebrations—was over. He looked at me as if he hadn't heard a thing. I didn't know how many times the words "over, over" tumbled from my lips, but they kept repeating in my mind like shots fired from a wurlitzer. He still didn't understand. My heart was breaking. I was too upset to communicate clearly.

"Let's spend the night together, like old times, so we can talk and you can learn what has happened." I offered to take him to the rest camp at Palmwag. We could visit there.

That evening, Archie told us he'd been questioned while we were in the United States.

He was asked: "Do Joel and Carol really shoot guns to see how far rhinos run?" and "Are there really hyenas in the Springbok River?"

"What did you tell them, Archie?"

"I told them the truth, Joel, that Carol shot the gun when *Kokerboom* came for us. That's all." Then he laughed, saying that it was funny that people didn't know that hyenas were in the Springbok River.

We became serious. "Joel, I just can't believe you are leaving. What will I do for work?" We looked long at each other. Archie broke the gaze, shifting to the ground. I finally spoke.

"Archie, I'll write and help as much as I can from America."

"I'll send my letters to you in Nevada. Will you be there? Do you think I can visit someday? I need help. I'm afraid to come alone."

We shook hands behind the Land Rover. A lump formed in my throat. Archie slipped me a "high five" and smiled. I did the same. Carol walked to the rear and they shook hands.

"Goodbye, Carol. Goodbye, Sonja."

Soni blew a kiss from her little hand. I pushed the clutch into gear. In the mirror stood Archie, pack in one hand, bedroll in the other. Extra mosquito netting and supplies he could use in Sesfontein lay on the ground. We accelerated away. Archie faded in the dust—an apparition, a presence, a part of our us.

We had learned much from someone straight from the core of Namibia, had tried to understand the world he experiences. We loved Archie.

At Okaukuejo we packed, leaving behind what we couldn't take on the plane. Our signed agreement with the ministry had stated that radiotelemetry and other equipment would be donated, so we took antennas, radiocollars, and receivers to the Institute. Finally organized, we headed to the airstrip for a last jog.

Carol scanned with her binoculars and found an unidentified lump next to a distant bush. *Nothing dangerous is going to be out here in this heat.* Even though it was 5:30 P.M., lions wouldn't be active at 95°F. We proceeded to the far end of the airstrip. For fun, Sonja helped steer. Carol refocused on the spot, yellow-brown, like a dirty lemon.

A male lion rested with its bloody kill. I looked, and it seemed far away. Besides, we really did need a jog and something pleasant to finish our Etosha stay. With one of us watching the lion, the other should be safe. We'd easily alert the other person. I began running.

Carol played with Sonja, looking up every minute or two. Then her heart sank. The lion was gone. Frantically, she told Sonja to play or read by herself, climbed onto the Land Rover, peering for any sign of the lion. She glanced back and forth. Nothing. She saw me, a tiny figure running steadily. Still no lion.

A vulture moved from the carcass. Then, a large head popped up. The lion had been there all along. A stork stood to one side.

I returned. Carol asked if I would watch as closely as she had. Then she handed me the book she had almost finished reading to Sonja, *Tails, Paws, Fangs, and Claws.*

Before passing through the final Etosha gate on our way to Windhoek, we stopped at the local school for a final goodbye, one as important to ending our Etosha life as Archie's had been to our time in the deserts. Elsie, Sonja's surrogate mom, came out. We hugged, knowing that this was it, the end; knowing that the distance between our countries made future face–to–face meetings unlikely. We climbed into the Land Rover to avoid more tears. As we pulled away, Sonja stuck her head out the window and, with her little voice, squeaked, "Elsie—I love you."

29

Xenophobia

[Carol]

Even during times of plenty, landowners view strangers with suspicion. Lions, ants, and humans are among the many species that exhibit xenophobia. There's disdain for the unfamiliar, for members of a different group, immigrants, foreigners. Kin are favored over nonkin.

Although not all societies are suspicious of foreigners, distrust is more the rule than the exception. Tadpoles prefer to be with close relatives. Lions generally form coalitions with pride mates. Among chimpanzees and humans group bonds are powerful, and tribal wars are legion.

The biological basis for such behavior has been strong. Individuals who favored known individuals lived to reproduce better than those who didn't. Hence, more of the former were represented in the next generation. The distrustful multiplied. Outsiders became increasingly unwelcome, particularly in closed social groups, human and nonhuman alike.

* * *

With Africa behind us, the politics of conservation still lay ahead. While tourists admired Namibia's tranquil settings and wildlife, the uneasy truce between foreign researchers and Namibian environmental officials soured. Although some outsiders continued studies within Namibia's borders, research that might affect the management of conservation areas or tactics was no longer continuing, despite contrary claims. *Africa Environment and Wildlife,* a respected journal from South Africa, carried a piece about Etosha's pan and wildlife, stating that its researchers studied "cheetah behaviour and the impact of this predator . . . the status of black-faced impala populations . . . [and] the social behavior . . . and the effects of dehorning [in rhinos]." We, of course, knew better.

None of the projects were actually by Namibians. Aron Rothstein and Wendy Green studied impalas. The Gasaways had been behind the cheetah and other predator projects. Then there was our rhino study. Both of the latter had been discontinued. An elephant project by an Oxford University student stalled midstream. A proposal to study elephants by a Kenyan was refused. Yet the sovereignty or rights of Namibians to control the research performed within their borders were indisputable, irrespective of whether species were considered Namibia's or the world's.

Meanwhile, reports of rhino calf mortalities found their way across the globe. The *New York Times, Nature, Australian Natural History,* the *Toronto Globe,* the *San Diego Tribune,* and others carried information about dehorning. A few mentioned the "termination" of our project.

Much of the coverage upset us. We had reported in the journal *Science* the fates of 10 calves, 3 of which had perished, 7 of which didn't, saying: "Despite admittedly restricted samples, the differences [in survivorship] are striking. . . . The data offer a first assessment of an empirically derived relation between horns and [calf] recruitment." Many of the reports on our work omitted mention of the small sample and simplistically claimed that dehorning was not working. Although we thought we had carefully reported both the positive and negative aspects of dehorning, we understood the Namibian distrust of foreigners and biased, superficial reporting. If media reports about Namibia's management policies were disapproving and resulted from our published and internal papers, then we shouldn't be surprised by the ministry's irritation with us. But we hadn't bargained on the failure to renew our research clearance, when we had delivered reports and clearly shown a desire to continue the study and help in other ways. Namibia's outrage by outside scrutiny of "their" rhinos had surprised us. We were naive.

Malan was quoted in *Nature* as saying that Joel and I "grossly overreacted" and displayed "hypersensitivity to criticism," and that Namibia "is more than just another study area or project for a foreign scientist, to be coldly dissected in foreign scientific journals." He accused us of communicating "by way of American scientific journals, claiming that pressure from grant sources to publish quickly superseded the courtesy of obtaining comment from their hosts."

We obviously believed otherwise. As independent scientists, we were obligated to dissect facts without partisan views. Namibian authorities had had more than half a year to comment. We felt betrayed. We had even submitted a paper to Malan's own symposium on Etosha more than a year earlier and published another in a different African journal. It was an outright lie to claim that we only communicated through foreign journals.

In a letter published in *Science,* we pointed out that Namibia "should be free to manage their own resources any way that they so please. The choice to accept or discard information is ultimately theirs alone." We also suggested that it will be impossible to validate declarations about the success or failure of different programs, because Namibia's self-avowed policy of stockpiling horns for later trade muddies the distinction between science and advocacy. The conflict of interest is evident. Once scientists are sanctioned or cannot publish results that contradict policy, all credibility is lost.

Throughout the rest of 1994 and into the next two years, claims and counterclaims appeared in scientific journals and throughout conservation circles. Some American scientists suggested that we weren't interested in conservation, arguing that had we been committed we would have worked more with the government and the NGOs. Save the Rhino Trust claimed no knowledge of calf deaths, or of one, or two, depending on which piece was read. A Zimbabwean scientist said that Joel's publications "are largely a vehicle for gratuitous, value-laden conservation philosophy."

Fortunately, others disagreed. Dr. Phoebe Barnard, from the Namibian Evolutionary Ecology Group, published a letter in the British journal *Trends in Ecology and Evolution* in defense of foreign researchers, indicating that knowledge gained by host countries benefits local scientists by developing a more mature scientific and education infrastructure. A move was even afoot by some Namibian scientists to overturn our "deportation," but the government quashed it. The president of the Society of Conservation Biology, Dr. Peter Brussard, wrote in the Society's newsletter about the importance of scientific knowledge in decisionmaking. He was roundly criticized for "Africa bashing" by those in search of political correctness. The *Rhino and Elephant Foundation* chose not to carry a piece on the debacle, apparently because it was too controversial.

The imbroglio moved from the U.S. embassy in Windhoek and the Namibian Consulate in Washington, D.C., to the U.S. State Department. It almost reached the vice president of the United States: Al Gore flew to South Africa for Nelson Mandela's presidential inauguration and then on to Namibia, where he met with ministry officials and viewed wildlife. The rhino fracas was one of 10 agenda items initially listed for discussion, in part because the US Agency for International Development funds had supported some of our work, and Namibian officials had accepted funds and equipment by approving the research. (Our troubles had worsened; our bank accounts there had even been frozen.) But the fiasco never made it to Gore's desk.

On our behalf, the U.S. embassy in Windhoek raised the possibility of our return to Namibia to continue research. Permanent Secretary Rumpf was intractable, adamant that we would not be allowed to return. Despite Malan's decree in *Nature* that we should have stayed because problems could have been worked out, we felt our decision to leave Namibia had been the right one. The permanent secretary's resolve could not be changed. Our rhino days were over.

* * *

Memories of the Africa we knew and loved will fade until that life seems a distant dream. Reminders linger—letters from friends, photographs, the incessant politi-

cal fallout. Salvaged mementos remain. A rhino, carved from leadwood by an An-golan, sits dusty on our living room floor. Framed above it are a small copper hose joint from our burned Land Rover and the darkened shell casing JB fired when the lioness charged. There is also a postage stamp of two masked Namibians, and a photo of a rhino at night—horns in darkness.

Epilogue

Although conservationists usually work toward common goals, perspectives on how to accomplish them frequently differ. Those of us born and educated in the crowded West, where industrialism and habitat conversion have erased many species and systems, hold convictions about the urgency of protection and the role of science in conservation. These views often contrast with those of people living in less developed countries, places where biodiversity has been under fewer assaults. A fundamental problem confounded us: How can "western" or "northern" scientists fit in, interact, and successfully share in the conservation vision when local research interests are defensive, territorial, or simply out of control?

The answers aren't easy. Conservation takes time. As when two enemies are suspicious of each other's motives, peaceful coexistence requires communication and understanding, mutual learning and trust. So does conservation, even if situations call for sitting under the outstretched arms of an acacia drinking homemade brew for six months with the headman, giving lectures at a local university, or writing administrative reports in stuffy government offices. There is no specific prescription to remove distrust.

* * *

It's ironic that the world is rapidly running out of space for its large mammals: Sufficient space does exist. But on a planet polarized between haves and have-nots, where wealth and education are unevenly distributed, and where women and other subgroups do not have equal access to opportunity, one would have to be naive to believe that the future of all living beings is rosy. Yet, despite our pessimism, there are reasons to remain optimistic.

Just last century, there were slaughters of wildlife as dramatic as the one that has lately so insidiously removed 97 percent of Africa's black rhinos. North American bison were once reduced to a few individuals in small reserves; now they number more than 150,000, with free-ranging populations in Canada, Alaska, and the Greater Yellowstone ecosystem. Perhaps rhinos too can make a comeback. Already, in India and Nepal, the greater one-horned rhino is at its highest level in half a century. The poaching of African rhinos that was so rampant during the 1980s and early 1990s is slowing. Most black and white rhinos are now sequestered in safe reserves, protected behind fences, with armed guards. We can hope their liberation is not too far off.

If there is anything to be learned, it is that the world has to unite to save its heritage, a biological legacy that involves more than rhinos. Conservation is the responsibility of all nations, not just a few. If we cannot protect rhinos, why should we expect a better fate for ourselves?

There is no perfect formula for conservation. What works in Kenya may not in South Africa or Zaire. But there must be incentives—ethical and moral, biological and economic. Local people must be involved and feel for themselves why conservation matters. It does little good for foreign instigators to argue that there are benefits from wildlife if local people receive nothing.

Inevitably, tactics will vary, locally, regionally, and internationally. In late 1995, the USFWS approved permits for darting safaris. Citizen darters will pay heavily for the privilege and work with professionals in host countries if they wish to return with horns in hand. "Trophy fees" will be $10,000 for the darting privilege and another $2,000 per kilogram for the horns. According to Zimbabwe, their main purpose in sanctioning the experience is to raise money for black rhino conservation.

Elsewhere, other strategies are being used. South Africa is reducing the risk of losing animals by beginning small "founder" populations in areas where suitable habitat still exists. Malawi guards their rhinos in small reserves. Censuses are underway in the Cameroons. Heavy protection has resulted in increases in Kenya's and Zimbabwe's black rhinos. Funding is what seems to work, money for guards and for educational programs.

Still, the world's last large population of free-roaming black rhinos is in Namibia. There—in the world's oldest desert, on stark gravel plains, and deep within canyons and remote mountains—rhinos travel thousands of kilometers, coursing dry rivers and sand to search for food and mates.

Despite humans' barbarism to one another and to nature, the twentieth century has not spelled total darkness for black rhinos. The Kaokoveld population has nearly doubled. The twenty-first century offers new opportunities. There is time for vision, a chance for victory.

Postscript

Eight months after we left Namibia, our bank funds were released to us. Malan was promoted to deputy director and lives in Windhoek.

* * *

Life goes on. After rhinos, we deliberately chose a less political project. Now, instead of waking to the smells of African savannas and the vision of antelopes leaping across dusty plains, we wander glacier-carved mountains and valleys in Alaska and the Greater Yellowstone area. We hope to learn more about ecosystems with and without grizzly bears and wolves.

Sonja, now seven, is adjusting to the new project and new places but, like her parents, still dreams of rhinos. As for Archie, he lives in Sesfontein but finds work, only sporadically as a guide, interpreter, and conservationist. His children are all in school. Perhaps, for a new generation there is hope.

Acknowledgments

Without the selfless commitment and untiring dedication of countless people and organizations, far fewer African rhinos would exist. We are particularly grateful to those who helped us: the Wildlife Conservation Society, the World Wildlife Fund (U.S.), the National Geographic Society, the American Philosophical Society, Rhino Rescue Ltd., the National Science Foundation, the United States Agency for International Development, the Hasselblad Foundation, Camenzind Productions, the Frankfurt Zoological Society, the Smithsonian Institution, and the University of Nevada. South African Air and Virgin Atlantic Airways waived charges more than once for excess gear.

In Zimbabwe, Drs. Michael Kock and Nancy Kock revitalized us with their encouragement and affirmed the importance of scientifically acquired data as the key to successful conservation management. In Namibia, Dr. Eugene Joubert has been a stalwart defender of rhinos since the 1960s. He took us under his wing and helped us gain access to data and rhinos. Duncan Gilchrist and Peter Erb began the Damaraland monitoring program. Blythe Loutit took over and promoted the efforts of Save the Rhino Trust. Ministry employees who helped in ways too numerous to elaborate include Dr. Nad Brain, Martin Britz, Raymond Dujardin, Wynand du Plessis, Mike Griffin, Dannie Grobler, Tommy Hall, Rudi Loutit, Dave Murray, Fritz Schenk, and Dr. Philip "Flip" Stander. Allan Cilliers, Kallie Venzke, and Lue Scheepers helped with photographic techniques, and later Kallie and Alan provided access to records on rhino mortalities and horn sizes. Lue taught us about lions and life in the veld. Dr. Pauline Lindeque was a true friend through difficult times and, together with Peter Erb, shared insights and made rhino files available. Dr. Betsy Fox put in long hours at the Okaukuejo water hole and contributed observations on interactions between lions and rhinos. Pierre du Preez and Duncan Gilchrist openly exchanged information and never wavered from the truth. From the private sector, help, discussion, and insight came from Peter Mostert, Dr. Joh Henschel, and Steve and Louise Braine. Land Rover specialists Luigi Ballotti and Herbie Klein in Windhoek and the Weimans in Outjo always found the time to deal with our vehicle and to take care of us as we waited. Francois and Linda Malan rescued us after our Land Rover burned.

From Dr. Pete Morkel and his wife Estelle we learned about life in the bush with little children. Pete is among the most committed field people we know. Without the backing of Eugene Joubert and Dr. Malan Lindeque this project would have never gotten off the ground. Malan was a supporter for several years. Mark and Dr. Delia Owens offered inspiration spanning nearly two decades.

Garth Owen-Smith and Margie Jacobsohn taught us about programs in community development and environmental education. They took us tracking, encouraged us, and offered critical observations. They opened their arms to us, sometimes to their detriment. Together, they strengthen the link between human and wildlife. Colin Nott and Tim and Rosie Holmes relayed sightings of desert rhinos and were true friends during trying times.

Archie Gawuseb was our connection to the desert. Without Archie, our data would be far fewer and our lives less rich. The times were wonderful and memorable; Archie, from the bottom of our hearts, thanks!

Wilfred and Elsie Versfeld provided sanity in a challenging world. From biltong to mechanics and watching your boys grow, we'll never forget.

Support from colleagues at the University of Nevada at Reno was steady, especially from Drs. Fred Gifford, Bernard Jones, and Ken Hunter. Janet Rachlow shared information, despair, and joys based on her experiences in Zimbabwe. Her knowledge about white rhino behavior and her challenges about conservation and policy consistently caused us to rethink our ideas.

At the Smithsonian Institution's Conservation and Research Center, Dr. Chris Wemmer provided a splendid environment in which to write. At the Wildlife Conservation Society, Drs. John Robinson, Amy Vedder, and Bill Weber never lost faith in us and the Society continues to support field conservation projects in Africa and elsewhere. Dr. Mary Pearl from the Wildlife Preservation Trust helped secure funding. At World Wildlife Fund, Drs. Eric Dinerstein and Cynthia Jensen were most helpful with ideas and expenses.

Dr. Franz Camenzind, along with Chip Houseman and Janie Camenzind, enriched our lives during five full months. They shared meals, Sonja-sitting chores, and charging rhinos. Franz has brought conservation projects into the living rooms of millions and has shaped our beliefs about the importance of activism and involvement.

Dr. Andy Phillips and Janet Wotherspoon introduced us to Namibian banking and Land Rovers and arranged for us to use the Zoological Society of San Diego's Caravan in Etosha. They also shared insights about leguuans, chemoreception, and deadly slangs. Andy graciously allowed use of his photos. Drs. Aron Rothstein and Wendy Green gave yearly updates and shared meals with us in deserts 20,000 kilometers apart. We now understand just how challenging black-faced impala can be.

Dr. Bill and Kathy Gasaway are proof that altruists exist. They devoted themselves to Namibian conservation issues. They introduced us to the African bush and their scrumptious campfire-baked bread. They tracked with us and shared their data, books, and ideas. We'll never forget sumps, aardvark holes, or the stony paths that we traveled together. Their friendship, on whatever continent, means a lot.

To the colleagues who helped shape our thoughts and analyses, Beaux Berkeley and Janet Rachlow, Drs. Peter Brussard, Eric Dinerstein, Bill Gasaway, Guy Hoelzer, Eugene Joubert, Bill Longland, Pete Stacey, and Gary White, we offer our most sincere thanks. The United States State Department and the National Science Foundation, especially Drs. Peter Thomas, Francis Li, and Mark Courtney, intervened on our behalf.

During the writing of this book, Scott Simons, Chip Rawlins, Terry Springer–Farley, and especially Kathy and Bill Gasaway made invaluable suggestions. Our editor, Kirk Jensen, was consistently upbeat, honing our manuscript and offering sage advice as we struggled forward.

Letters from our families and care packages filled with news clippings, audio cassettes, books, and food made memories of America less distant and those of parents and sibs more alive. Robert Gibson and Gwen Bachman sweetened our disposition from afar with very special chocolates.

Drs. Rob Simmons and Phoebe Barnard are courageous allies and companions in the battle for conservation. They shared—food, shelter, and scientific camaraderie. Together, they embody the true practitioner, those who toil in government and nongovernment offices and who maintain courage when victories are so few. If more people were like them, the world would be a better place for human and nonhuman alike.

Glossary of Names, Terms, and Key Characters

Afrikaner a descendant of Dutch settlers of South Africa
Afrikaans language spoken by Afrikaners
bakkie small pick-up truck
borehole human-made well or windmill
boma enclosure built with wooden poles
caravan trailer or portable housing unit towed by a vehicle
Damara name of group of people in western Namibia
Damaraland region occupied by the Damara people; it is the desert region of the southern Kunene Province, also known as the southern Kaokoveld
fountain a spring or waterhole
Herero name of a group of people in Namibia, southern Angola, and western Botswana
Kaokoveld region of the northern Namib Desert, within the Kunene Province
Kunene Province a region of northwestern Namibia
leguaan a type of savanna monitor lizard
mealie corn meal
ministry the Ministry of Wildlife, Conservation and Tourism, now called the Ministry of Environment and Tourism
!nabis Damara word for rhino
Okaukuejo major tourist camp in south central Etosha National Park
ongava Herero word for rhino
pan open areas that may collect water seasonally and then dry out
panga wide-bladed knife, similar to a machete
slang Afrikaans word for snake
shifta Somalian bandits
spoor animal or vehicle tracks, animal sign or feces, small dirt road or track
tarmac paved road
vlei open areas, usually within forested regions

Key Characters

Barnard, Phoebe biologist in Windhoek, and working in Namibian National Biodiversity Programme (ministry)
Camenzind, Franz biologist turned filmmaker and producer
Camenzind, Janie sound recordist, photographer

Cilliers, Allan head of management (ministry) staff at Etosha; joined private sector

Dujardin, Raymond Chief ranger of Halali District (ministry) in Etosha; retired

du Preez, Pierre government ranger of Outjovassandu District (ministry) in Etosha, moved to Caprivi

Du Toit, Raoul biologist in Zimbabwe

Erb, Peter ministry biologist at Waterberg Plateau Park; moved to Etosha

Gasaway, Bill biologist in Etosha; moved to Utah

Gasaway, Kathy researcher with Bill; moved to Utah

Gawuseb, Archie known as !Huie, tracker with our project

Geldenhuys, Louis Leader of Game Capture Unit (ministry); charged with animal theft and currently in private sector

Green, Wendy biologist studying black-faced impala through the University of Nevada

Grobler, Dannie head of management (ministry) in Windhoek

!Huie the Damara name for Archie Gawuseb

Houseman, Chip filmmaker

Jacobsohn, Margaret anthropologist with Integrated Rural Community Development

Joubert, Eugene head of research (ministry) in Windhoek; moved to Saudi Arabia

Lindeque, Malan director of research (ministry) in Etosha; now a deputy director in Windhoek

Lindeque, Pauline ministry biologist in Etosha; now a biologist (ministry) in Windhoek

Loutit, Blythe director of field operations for Save the Rhino Trust in Khorixas

Loutit, Rudi chief conservator (ministry) for Damaraland; now studying for a master's degree in Queensland, Australia

Montgomery, Sharon director of education for Save the Rhino Trust in Windhoek; resigned and writing freelance

Morkel, Pete veterinarian (ministry) at Etosha; joined private sector and later moved to South Africa

Murray, Dave chief of the Anti-Poaching Unit (ministry) in Etosha; joined private sector

Owen-Smith, Garth Director of Integrated Rural Community Development

Pelser, Piet Chief of Anti-Poaching Unit (ministry) in Etosha; deceased

Phillips, J. A. ("Andy") biologist of the Zoological Society of San Diego

Rachlow, Janet University of Nevada doctoral candidate studying white rhinos in Zimbabwe

Riley, Mickey government ranger of Outjovassandu District (ministry) in Etosha; joined private sector

Rothstein, Aron biologist studying black-faced impala through University of Nevada

Rumpf, Hanno permanent secretary of the ministry in Windhoek; now permanent secretary of the ministry of Trade and Industry

Scheepers, Lue ministry biologist in Etosha; transferred to Katima Muhlilo

Schraeder, Henti acting head of research (ministry) Windhoek; joined private sector

Simmons, Rob ornithologist (ministry) in Windhoek

Swart, Polla director of the ministry in Windhoek; retired

Tjimbura, Muhinene headman of small group of Himas in the northern Kaokoveld

Ulenga, Ben deputy minister of the ministry in Windhoek; transferred to the ministry of Regional and Local Government and Housing

Venzke, Kallie head of management (ministry) in Etosha

Versfeld, Wilfred ministry biotechnician in Etosha

Selected Bibliography

Sources

Adams, J. S. and T. O. McShane. 1992. *The myth of wild Africa; conservation without illusion*. W. W. Norton, New York.

Akeley, C. E. 1925. *In brightest Africa*. Garden City Publishing, Garden City, New York.

Alexander, J. E. 1838. *An expedition of discovery into the interior of Africa*. Two vols. Henry Colburn, London.

Anderson, D., and R. Grove. 1987. *Conservation in Africa; people, policies, and practice*. Cambridge University Press, Cambridge, England.

Andersson, C. J. 1856. *Lake Ngami; or, explorations and discoveries, during four years of wanderings in the wilds of South Western Africa*. Hurst and Blackett, London.

Anonymous, 1992. Zimbabwe black rhino conservation strategy. Zimbabwe Department of National Parks and Wild Life Management, Harare.

Barnard, P. 1995. Scientific traditions and collaboration in tropical ecology. *Trends in Ecology and Evolution* 10:38–39.

Bartlett, D., and J. Bartlett. 1992. Africa's Skeleton Coast. *National Geographic* 181: 55–85.

Bonner, R. 1993. *At the hand of man; peril and hope for Africa's wildlife*. Knopf, New York.

Booth, M. 1992. *Rhino road; the black and white rhinos of Africa*. Constable, London.

Borner, M. 1981. Black rhino disaster in Tanzania. *Oryx* 16:59–66.

Bradley-Martin, E., and C. Bradley-Martin. 1982. *Run rhino run*. Chatto and Windus, London.

Brandon, H. 1993. The mopane worm—an unlikely source of protein. *Africa Environment and Wildlife* 1:77–79.

Chapman, J. 1868. *Travels in the interior of South Africa, 1849–1863*. (Reprinted 1971.) A. A. Blakema, Cape Town.

Curry-Lindahl, K. 1970. War and the white rhino. *Oryx* 4:263–267.

Darwin, C. 1871. *The descent of man in relation to sex*. [Murray Reprint], New York (Hurst, 1874)

Delegorgue, A. 1847. *Travels in southern Africa*. (Reprint.) University of Natal Press, Pietermaritzburg.

Douglas-Hamilton, I., and O. Douglas-Hamilton. 1992. *Battle for the elephants*. Transworld, London.

Edroma, E. L. 1982. White rhino extinct in Uganda. *Oryx* 16:352–355.

Estes, R. D. 1991. *The behavior guide to African mammals.* University of California Press, Berkeley.

Gasaway, W. C., Gasaway, K. T., and Berry, H. H. 1996. Persistent low densities of plains ungulates in Etosha National Park, Namibia: testing the food regulating hypothesis. *Canadian Journal of Zoology* 74:1556–1572.

Ghazi, P. 1991. Save the rhino. *Observer,* September 1:9–14.

Goddard, J. 1967. Home range, behaviour and recruitment rates of two black rhinoceros populations. *East African Wildlife Journal* 5:133–150.

Goddard, J. 1969. A note on the absence of pinnae in the black rhinoceros. *East African Wildlife Journal* 7:179–180.

Green, F. 1860. In: Charles John Anderson papers, Vol 2. *Archeia* (State Museum of Namibia) 10; Windhoek.

Green, L. G. 1952. *Lords of the last frontier; the story of South West Africa and its people of all races.* Howard B. Timmins, Cape Town.

Guggisberg, C. A. W. 1966. *S. O. S. rhino.* Trinity Press, Worcester, England.

Hall-Martin, A., C. Walker, and J. du P. Bothma. 1988. *Kaokoveld; the last wilderness.* Southern Book Publishers, Johannesburg.

Hamilton, W. J. III, R. Buskirk, and W. H. Buskirk. 1977. Intersexual dominance and differential mortality of gemsbok *Oryx gazella* at Namib Desert waterholes. *Madoqua* 10:5–19.

Hitchins, P. A. 1986. Earlessness in the black rhinoceros—a warning. *Pachyderm* 7: 8–10.

Holstenson, M. 1993. The Caprivi Strip: history, geography, and legend. *Endangered Wildlife* 15:15–17.

Huntley, B. 1972. An eden called Iona. *African Wildlife* 26:136–141.

Imperato, J. P. and M. E. Imperato. 1992. *They Married Adventure: The Wandering Lives of Martin and Osa Johnson.* Rutgers University Press, New Brunswick.

Jacobsohn, M. 1990. *Himba; nomads of Namibia.* Struik Publishers, Cape Town.

Johns, M. 1994. The great white place; Namibia's Etosha National Park, Africa. *Environment and Wildlife* 2:61–69.

Jones, B. 1987. Death of S. W. A. Namibia's coastal lions. *African Wildlife* 41:297.

Joubert, E. 1971. The past distribution and present status of the black rhinoceros (*Diceros bicornis*) in South West Africa. *Madoqua* 1:33–43.

Joubert, E. 1971. Observations on the habitat preferences and population dynamics of the black-faced impala *Aepfceros petersi* Bocage, 1875, in South West Africa. *Madoqua* 1:55–65.

Kasale, B. and R. Pakleppa. 1991. Namibia, the land hungry people. *New Ground* 1: 20–23.

Largen, M. J., and D. W. Yalden. 1987. The decline of elephant and black rhinoceros in Ethiopia. *Oryx* 21:103–106.

Lau, B. 1989. *Charles John Anderrson; trade and politics in central Namibia 1860–1864. Archeia* (State Museum of Namibia), Windhoek.

Leader-Williams, N. 1992. *The world trade in rhino horn: a review.* Traffic International, Cambridge, England.

Liebenberg, L. 1990. *The art of tracking; the origin of science.* David Phillip Publishers, Claremont, South Africa.

Lindeque, M. 1990. The case for dehorning the black rhinoceros in Namibia. *South African Journal of Science* 86:226–227.

Marsh, A. C. 1993. Ants that are not too hot to trot. *Natural History* 102:43–44.

Martin, H. 1957. *The sheltering desert.* William Kimber, London.

Merrett, P. L., and R. Butcher. 1991. *Robert Briggs Struthers hunting journal 1852–1856 in the Zulu kingdom and the Tsonga regions.* University of Natal Press, Pietermaritzburg.

Milliken, T., E. B. Martin, and K. Nowell. 1991. Rhino horn trade controls in East Asia. *Traffic Bulletin* 12:17–21.

Mills, M. G. L. 1990. *Kalahari hyenas; the comparative behavioural ecology of two species.* Unwin Hyman, London.

Milner-Gulland, E. J., J. R. Beddington, and N. Leader-Williams. 1992. Is dehorning African rhinos worthwhile? *Pachyderm* 17:52–58.

Moller, P. 1899. *Resa I Afrika genom Angola, Ovampo och Damaraland.* Wilhelm Billes Bokforlags Aktiebolag, Stockholm.

Neihardt, J. G. 1961. *Black Elk Speaks.* William Morrow and Co., New York

Owens, M., and D. Owens. 1984. *Cry of the Kalahari.* Houghton-Mifflin, Boston.

Owen-Smith, G. 1986. The Kaokoveld, South West Africa/Namibia's threatened wilderness. *African Wildlife* 40:104–115.

Owen-Smith, R. N. 1988. *Megaherbivores; the influence of very large body size on ecology.* Cambridge University Press, Cambridge, England.

Phillips, J. A. 1995. Movement patterns and density of *Varanus albigularis. Journal of Herpetology* 29:407–416.

Pradervand, P. 1989. *Listening to Africa; developing Africa from the grassroots.* Praeger, New York.

Raath, J. 1992. Giant pig with number on back. *Africa South* (March):40–41.

Rookmaaker, L. C. 1989. *The zoological exploration of southern Africa 1650–1790.* A. A. Balkkema, Rotterdam.

Rosenblum, M. and D. Williamson. 1987. *Squandering Eden.* Harcourt Brace Jovanovich, San Diego.

Ryder, O. A. 1993. *Rhinoceros biology and conservation.* San Diego Zoological Society, San Diego.

Seely, M. 1987. *The Namib; natural history of an ancient desert.* Shell Oil Ltd., Windhoek.

Selous, F. C. 1908. *Africa nature notes and reminiscences.* MacMillan, London.

Serton, P. 1954. *The narrative journal of Gerald McKiernan in South West Africa, 1874–1879.* Rustica Press, Wynberg, Cape Town.

Spinage, C. A. 1986. The rhinos of the Central African Republic. *Pachyderm* 6:10–13.

Stanley, H. M. 1890. *In darkest Africa.* Scribner's, New York.

Tudge, C. 1991. Can we end rhino poaching? *New Scientist* (October 5):34–39.

Van der Merwe, C. 1989. Spiking the guns. *Leadership* 8:84–95.

Vigne, L. 1989. Dehorning rhinos in Damaraland—a controversial issue. (Excerpts from B. Loutit, Save the Rhino Trust Fund Newsletter 53). *Pachyderm* 12:47–48.

Western, D. 1982. Patterns of depletion in a Kenya rhino population and the conservation implications. *Biological Conservation* 24:147–156.

Western, D. 1987. Africa's elephants and rhinos: flagships in crisis. *Trends in Ecology and Evolution* 2:343–346.

Scientific Publications Resulting from Our Work

Berger, J. 1993. Rhino conservation tactics. *Nature* 361:121.

Berger, J. 1993. Disassociations between black rhino mothers and young calves: ecologically variable or, as yet, undetected behavior? *African Journal of Ecology* 31: 261–264.

Berger, J., C. Cunningham, A. Gawuseb, and M. Lindeque. 1993. "Costs" and short-term survivorship of hornless black rhinos. *Conservation Biology* 7:920–924.

Berger, J. 1994. Science, conservation, and black rhinos. *Journal of Mammalogy* 75: 298–308.

Berger, J., and C. Cunningham. 1994. Active intervention and conservation: Africa's pachyderm problem. *Science* 263:1241–1242.

Berger, J., and C. Cunningham. 1994. Phenotypic alterations, evolutionary significant structures, and rhino conservation. *Conservation Biology* 8:833–840.

Berger, J., and C. Cunningham. 1994. Black rhino conservation. *Science* 264:757.

Berger, J., and C. Cunningham. 1994. Horns, hyenas, and black rhinos. *Research and Exploration* 10:241–244.

Berger, J., C. Cunningham, and A. A. Gawuseb. 1994. The uncertainty of data and dehorning black rhinos. *Conservation Biology* 8:1149–1152.

Berger, J., and C. Cunningham. 1995. Predation, sensitivity, and sex; why female black rhinoceroses outlive males. *Behavioral Ecology* 9:57–64.

Berger, J., and C. Cunningham. 1996: Is rhino dehorning scientifically prudent. *Pachyderm* 21:60–68.

Berger, J. 1996. Animal behaviour and plundered mammals: Is the study of mating systems a scientific luxury or a conservation necessity? *Oikos* 77: in press

Berger, J. 1997. Population constraints associated with the use of black rhinos as an umbrella species for desert herbivores. *Conservation Biology* 11: in press

Brussard, P. F. 1994. Science and conservation agencies: still an uneasy partnership. *Conservation Biology Newsletter* 1:1.

Lindeque, M., and K. P. Erb, 1995. Research on the effects of temporary horn removal on black rhinos in Namibia. *Pachyderm* 20:27–30.

Loutit, B., and S. Montgomery. 1994. Rhino conservation. *Science* 265:1157–1158.

Loutit, B., and S. Montgomery. 1994. The efficacy of rhino dehorning: Too early to tell! *Conservation Biology* 8:923–924.

Macilwain, C. 1994. Biologists out of Africa over rhino dispute. *Nature* 368:677.

Rachlow, J., and J. Berger. 1997. Conservation implications of horn regeneration in white rhinos. *Conservation Biology* (In press)

Index